GERMAN SELF-PROPELLED ARTILLERY GUNS OF THE SECOND WORLD WAR

I0894479

GERMAN SELF-PROPELLED ARTILLERY GUNS OF THE SECOND WORLD WAR

CRAIG MOORE

FONTHILL

Fonthill Media Language Policy

Fonthill Media publishes in the international English language market. One language edition is published worldwide. As there are minor differences in spelling and presentation, especially with regard to American English and British English, a policy is necessary to define which form of English to use. The Fonthill Policy is to use the form of English native to the author. Craig Moore was born and educated in London; therefore, British English has been adopted in this publication.

Fonthill Media Limited
Fonthill Media LLC
www.fonthillmedia.com
office@fonthillmedia.com

First published in the United Kingdom and the United States of America 2019

British Library Cataloguing in Publication Data:
A catalogue record for this book is available from the British Library

ISBN 978-1-78155-695-5

Typeset in 10pt on 13pt Sabon
Printed and bound by CPI Group (UK) Ltd, Croydon, CR0 4YY

Acknowledgements

Special thanks go to David Bocquelet for providing the illustrations.

I would like to thank the following people who have helped me in writing this book: Herbert Ackermans, Jason Belgrave, Jon Bernstein, David Bocquelet, Pierre-Oliver Buan, Marcus Hock, Tim Issac, Pascale Mathieu, Stephen MacHaye, Christophe Mialon, Steve Osfield, Yuri Pasholok, Ralf Raths, and Lucian Stan.

I would also like to thank the following organisations: Armourgeddon Museum, Husbands Bosworth, Leicestershire, England; Arsenalen, Swedish Tank Museum, 645 91 Strängnäs, Sweden; Artillery, Engineer and Signals Museum, Hämeenlinna, Finland; Australian Armour and artillery Museum, Cairns, Australia; Central Museum of the Great Patriotic War 1941–1945, Park Pobedy, Moscow, Russia; Cobbaton Combat Collection, North Devon, England; Deutsches Panzermuseum, German Tank Museum, Munster, Germany; IWM Duxford, Cambridgeshire, England; Musée des Blindés, French Tank Museum, Saumur, France; Norfolk Tank Museum, Forncet St Peter, England; Royal Museum of the Armed Forces and of Military History, Brussels, Belgium; and The Tank Museum, Bovington, Dorset, England.

Contents

Glossary of Military Terms

Abteilung (Abt): Battalion; depending on its usage, this term could mean detachment, department, or battalion. The vast majority of the time, *Abteilung* meant battalion.
Alkett: German engineering and manufacturing company.
Antenne: aerial.
AP: armour-piercing.
Artillerie: artillery.
Auf: on.
Ausführung (Ausf.): version, series, model, type (normally followed by a letter).
Batterie (plural: *Batterien*): Battery—an artillery unit of three to six towed guns, self-propelled guns, or anti-aircraft weapons.
Baukommando: construction command.
Beutepanzer: trophy tanks.
Division (plural: *Divisionen*): division—an organisational unit generally made up of two or three regiments (*Regimenter*).
Dunkelgelb: dark sandy yellow tank paint colour.
Entwicklungsstück: prototype development piece.
Fahrgestellnummer (Fgst.Nr.): chassis.
Fallschirmjäger: paratroop units.
Feld-Versuchs Batterie: field test battery.
Flammenwerfer (Flamm): flamethrower.
Forges & Chantiers de la Méditerrannée (FCM): a French shipbuilder of Toulon who also built tanks.
Funk (Fu): radio.
Funkgerät (FuG): radio device.
Gebirgs-Artillerie: mountain artillery.
Gebirgshaubitze: mountain howitzer.
Gebirgsraupenschlepper (RSG): mountain caterpillar tractor.
Gepanzerte Zugkraftwagen (gep.Zgkw): armoured half-track.
Gepanzerte: armoured (as in *Gepanzerte Artillerie-Regiment*: armoured artillery regiment).
Geschütz: gun.
Geschützpanzer (Gp): gun tank or self-propelled gun.

Geschützwagen (GW): gun vehicle.
Grille: cricket.
HE: high-explosive.
Heer: the regular German Army.
Heereswaffenamt (HWA): German Army High Command.
Heuschrecke: grasshopper.
Hummel: bumblebee.
Infanterie: infantry.
Infanteriegeschütz: infantry gun.
Inspektorat 6 (In 6): army inspectors for armoured and motorised units.
Jagd: hunt, hunting; a *Jagdtiger* was the special tank hunter/ambush version of the Tiger tank.
Jagdpanzer: hunting tank.
Jäger: hunter; *Panzerjäger* means tank hunter.
Kampfgruppe: battle/combat group.
Kanone: cannon.
Kettenkrad: tracked motorcycle.
Klein(*e*): small.
Kommando (plural: *Kommandos*): a command in the sense of a geographical area of authority; not to be confused with the British Army Commando soldier.
Kompanie: company.
Korps: corps.
Kriegsmarine: the German Navy.
Krupp: German engineering and manufacturing company.
leFH (*leichte FeldHaubitze*): light field howitzer.
Leicht(*e*): light.
Leuchtgeschoss: illuminating star shell, a flare.
Luftwaffe: the German Air Force.
Maybach: German engine manufacturer.
Mörser: mortar.
Mörserlafette: mortar gun carriage.
Nebelwerfer: rocket artillery.
Panzer: armour or armoured; refers to German tanks and tank units.
Panzerabwehrkanone (PaK): anti-tank gun.
Panzerfeldhaubitze: armoured field howitzer.
Panzergranate (Pzgr): armour-piercing shell.
Panzergrau: tank grey paint colour.
Panzerjäger: anti-tank (literally means tank hunter).
Panzerkampfwagen (Pz.Kpfw): tank (armoured combat vehicle).
Raupenschlepper: tracked tractor.
Rheinmetall-Borsig: German engineering company.
Röhrenpulver: hollow tube explosive.
Rohrrückholer (Recuperator): a device employing springs or pneumatic power used to control the gun's recoil and to return a gun to the firing position after firing.

Schlepper: tractor.
Schnell(e): quick.
Schwer(e): heavy.
Schwere Feldhaubitze (sFH): heavy field howitzer.
Schwere Infanteriegeschütz (sIG): heavy infantry gun.
Selbstfahrlafette (Sf): self-propelled gun.
Sonderkraftfahrzeug (Sd.Kfz.): special purpose vehicle. Each vehicle accepted into German Army service was given an ordnance inventory designation.
SPG: self-propelled gun.
Sturm: assault.
Sturmartillerie: assault gun (literally assault artillery, the early-war term for the assault guns). Assault guns were part of the artillery branch and not armoured troops.
Sturmgeschütz: assault gun.
Sturmgeschütz-Abteilung (Stu.Gesch.Abt): assault gun battalion.
Sturmpanzer: self-propelled assault guns with fully enclosed armoured fighting compartment.
Versuch: trial or prototype.
Waffenamt (WaA): German Military Ordnance Department.
Waffen-SS: armed *Schutzstaffel*. Units of the political organisation of the SS, the German *Schutzstaffel*: The Waffen-SS is often mistaken for the SS itself, and although a part of the larger structure of the political SS, the Waffen-SS was a front-line fighting organisation.
Waffenträger: weapons carrier.
Wehrmacht: armed military forces.
Werfer: launcher or thrower.
Wespe: wasp.
Zur besonderen Verwendung (z.b.V.): for special deployment or for special assignment.

1
German Artillery

The German Army in the Second World War had three main types of artillery: anti-tank guns, anti-aircraft guns, and field howitzers that fired high-explosive (HE) fragmentation shells. This book concentrates on the latter and the development of self-propelled artillery guns (SPG).

On 14 October 1933, Germany left the League of Nations and the Disarmament Conference. The country started to rearm. On 1 October 1934, the first artillery regiment officially entered service. More followed. They consisted of four battalions. The fourth was normally the heavy artillery battalion.

Under the terms of the armistice following the First World War, Germany was not allowed to have a large modern army. Development of new artillery pieces was therefore carried out in secret during the 1920s and early 1930s.

The model number of these new guns was chosen to deceive any Allied weapons inspectors. The number '18' would infer that the gun was developed in 1918 during the previous war and thus was an old artillery piece (the 10.5-cm *leichte Feldhaubitze* 18 entered service in 1933 not 1918). The 15-cm sFH 13 had been developed in 1913 and the 10.5-cm leFH 16 was a First World War gun designed in 1916.

Field artillery howitzers are also called guns or cannons (in German: *Feldhaubitze*, *Geschütz*, or *Kanone*). These weapons are also subdivided by weight and use—light (*leichte*), heavy (*schwere*), infantry (*Infanterie*), mountain (*Gebirgsjäger*), or airborne (*Fallschirmjäger*) guns.

Most German infantry divisions had their own field artillery regiments (*Feldartillerie-Regimenter*). They were organised in batteries that contained either towed 10.5-cm *leichte Feldhaubitze* 18 (10.5-cm light field howitzers, often shortened to 10.5-cm leFH 18) or towed 15-cm *schwere Feldhaubitze* (15-cm heavy field howitzers, often shortened to 15-cm sFH 18). Other guns were also used.

Types of Artillery Gun Laying

There are two main methods of artillery gun laying: direct fire or indirect fire.

For direct fire, the artillery piece is sighted by the gun crew on a target that must be visible to the gunner. The gun is adjusted for direction and elevation, taking into consideration the effects of wind direction and air temperature. This type of firing method would be used to destroy an enemy-fortified farmhouse that could be seen in the distance.

For indirect fire, the gun crew cannot see their intended target. A map grid reference would be used to plot the target area. The target could be over 8 km (4.97 miles) away. Troops on the front line would radio back the location of the enemy to the battery commander requesting supporting fire.

Pre-planned indirect fire was often used to rain down high-explosive fragmentation shells on the enemy ahead of an attack in a rolling barrage that advanced forward a certain distance as the attack progressed.

Indirect fire is easily the predominant method used by artillery gun crews. Traditional towed guns and self-propelled artillery guns can fire effectively from protected and concealed positions in relative safety some distance from the front line.

Separate-Loading Two-Part Cased Charge

The high-explosive shell was in two pieces. It was a separate-loading two-part round. First, the high-explosive projectile would be loaded and then the cartridge propellant case. Depending on the range of the target, different sized bags of propellant were inserted into the cartridge. More bags were used for longer range targets.

The 10.5-cm *leichte Feldhaubitze* 16 (10.5-cm leFH 16)

The 10.5-cm leFH 16 gun was a German light field howitzer used in the First World War. It was mounted on the 10.5-cm leFH 16 (Sf) *auf Geschützwagen* MkVI(e) and 10.5-cm leFH 16 (Sf) *auf Geschützwagen* FCM(f) self-propelled artillery guns.

It had a shorter range than the Second World War 10.5-cm leFH 18 gun. As it had the same calibre as the newer leFH 18, it could fire the same ammunition.

The abbreviation leFH stands for the German words '*leichte Feldhaubitze*' which translated means 'light field howitzer'. The number '16' refers to 1916, the year it was introduced into the Imperial German Army. They were produced by the German weapons manufacturer Krupp during the First World War and designed as a 'partner piece' for the 7.7-cm FK 16 gun. They both used the same gun carriage, which helped with supplying spare parts.

When mounted on the 10.5-cm leFH 16 (Sf) *auf Geschützwagen* FCM(f) SPG, it was fitted with a '*Mündungbremse*' muzzle brake to allow longer range charges to be fired and reduce the amount of recoil on the gun. This increased the operational life of the gun barrel.

The 10.5-cm high-explosive fragmentation shell weighed 14.81 kg (32 lb 10 oz). Using charge bag number five, it had a maximum muzzle velocity of 395 mps (1,300 fps) and a maximum firing range of 9.22 km (5.73 miles). With a good gun crew, it had a rate of fire between four and five rounds per minute.

It could fire a lighter 12.3-kg (27-lb 2-oz) *Hohlladung* (HL) hollow charge shell. If a charge bag number six was used, it could fire this shell 1.5 km (0.93 miles) at a velocity of 405 mps (1,329 fps).

German 15-cm artillery cased cartridge cases into which bags of explosive propellant were placed before having the cap placed on the top and shoved into the gun breach after the high explosive shell had been loaded. On display at Tim Issac's Cobbaton Combat Collection, North Devon, England.

This 10.5-cm *leichte Feldhaubitze* (light field howitzer) 16 (10.5-cm leFH 16) is on display at the Imperial War Museum Duxford, Cambridgeshire, England.

This 10.5-cm *leichte Feldhaubitze* 16 (10.5-cm leFH 16) is on display at the Australian Armour and Artillery Museum, Cairns, Queensland. It is fitted with air-filled rubber tyres. (*Jason Belgrave AAAM*)

The 10.5-cm *leichte Feldhaubitze* 16 gun was not very useful in the direct-fire mode against enemy armoured vehicles. It could only penetrate 52 mm (2 inches) of armour plate at a very short range of 500 m. For self-defence, it could fire an armour-piercing shell that weighed 14.25 kg (31 lb 6 oz).

The high-explosive shell was in two pieces. It was a 'separate-loading' or two-part round: first, the projectile would be loaded and then the cartridge propellant case containing a charge bag.

It needed a team of six horses to tow it in three pairs. As well as the six-man gun crew required to operate the artillery piece, three additional men were necessary to manage the horses.

Specifications

Calibre: 105 mm (4.13 inches).
Length of gun: 2.31 m (7 feet 7 inches).
Length of bore: 1.88 m (6 feet 2 inches).
Rifling: thirty-two grooves, right-hand increasing twist, 1/45 to 1/18.
Breech mechanism: horizontal sliding block, percussion fired.
Traverse: 4 degrees.
Elevation: -10 to +40 degrees.
Weight in action: 1.45 tonnes (1.42 tons).

Maximum Firing Range

Firing standard high-explosive shell weighing 14.81 kg (32 lb 10 oz):
Charge one: velocity 199 mps (653 fps), maximum range 3.45 km (2.14 miles).
Charge two: velocity 233 mps (764 fps), maximum range 4.55 km (2.82 miles).
Charge three: velocity 265 mps (869 fps), maximum range 5.70 km (3.54 miles).
Charge four: velocity 320 mps (1,050 fps), maximum range 7.42 km (4.61 miles).
Charge five: velocity 395 mps (1,296 fps), maximum range 9.22 km (5.72 miles).

The 10.5-cm *leichte Feldhaubitze* 18 (10.5-cm leFH 18)

This gun was the standard divisional towed light field artillery howitzer of the German Army. It was mounted on the following self-propelled artillery guns:

10.5-cm leFH 18/1 (Sf) *auf Geschützwagen* IVb (Sd.Kfz.165/1)
10.5-cm leFH 18/3 (Sf) *auf Geschützwagen* B2(f)
10.5-cm leFH 18/2 (Sf) *auf Geschützwagen* II *Wespe* (Sd.Kfz.124)
10.5-cm leFH 18/6 *auf Waffenträger Geschützwagen* III/IV *Heuschrecke* IVb Grasshopper
10.5-cm leFH 18/40/2 (Sf) *auf Geschützwagen* III/IV
10.5-cm leFH-18/4 (Sf) *auf Geschützwagen* Lorraine Schlepper(f)
10.5-cm leFH 18 (Sf) *auf Geschützwagen* 39H(f)
10.5-cm leFH 18/2 (Sf) *auf Geschützwagen* III/IV Hummel-Wespe

The gun was designed by the German arms manufacturer Rheinmetall-Borsig AG (Rheinische Metallwaren und Maschinenfabrik) in 1929. It was the replacement for the 10.5-cm leFH 16. It entered service in 1933 and soon became the standard artillery piece for the Army.

It was not a radically new design. It had split-trail legs at the rear with folding spades on the end that could be dug in the ground to help with recoil. It had a heavy but simply designed breech mechanism.

The recoil and recuperator system is of the hydropneumatic type. The buffer was contained in a cradle below the gun barrel and the hydropneumatic recuperator was fitted above it. Underneath the cradle, there was a single hydropneumatic balancing press. This helped counterbalance the weight of the gun and thus lighten the gunner's task.

The towed version of the howitzer had a different gun carriage, with wooden-spoked wheels. It needed a team of six horses to tow it in three pairs. As well as the six-man gun crew needed to operate the artillery piece, three additional men were required to manage the horses. Traditional towed artillery guns were very manpower- and resource-intensive.

This gun was primary an artillery weapon that fired high-explosive fragmentation shells. In self-defence, it could fire anti-tank armour-piercing shells over short ranges. The 10-cm Pzgr AP shell weighed 14.25 kg (31 lb 7 oz). It contained 650 gm (1.43 lb) of TNT explosive. Just like the HE shells, it was two-part ammunition. A separate charge five bag was loaded after the AP shell had been rammed into the gun breech. This gave the shell a

This 10.5-cm *leichte Feldhaubitze* 18 (10.5-cm leFH 18) is on display at in the Armourgeddon Museum, Husbands Bosworth, Leicestershire, England.

velocity of 395 mps (1,296 fps). The nearer the enemy tank was increased the chances of knocking it out. This AP shell had a maximum effective range of 1.5 km (0.93 miles)

This is the armour penetration figures for the 10-cm Pzgr AP shell fired from a 10.5-cm leFH 18 gun: 500-m range: 67 mm (2.64 inches), armour plate angled at 30 degrees, 56 mm (2.20 inches); 1,000-m range: 62 mm (2.44 inches), armour plate angled at 30 degrees, 52 mm (2.05 inches); and 1,500-m range: 59 mm (2.32 inches), armour plate angled at 30 degrees, 49 mm (1.93 inches).

Specifications

Calibre: 105 mm (4.13 inches).
Length of gun: 2.94 m (9 feet 8 inches).
Length of bore: 2.61 m (8 feet 7 inches).
Rifling: 32 grooves, right-hand increasing twist, 1/30 to 1/15.
Breech mechanism: horizontal sliding block, percussion fired.
Traverse: 56 degrees.
Elevation: –6 degrees 30 minutes to +40 degrees 30 minutes.
Weight in action: 1.98 tonnes (1.95 tons).

Maximum Firing Range

Firing standard high-explosive shell weighing 14.81 kg (32 lb 10 oz):
Charge one: velocity 200 mps (656 fps), maximum range 3.57 km (2.21 miles).
Charge two: velocity 232 mps (761 fps), maximum range 4.62 km (2.87 miles).
Charge three: velocity 264 mps (866 fps), maximum range 5.76 km (3.57 miles).
Charge four: velocity 317 mps (1,040 fps), maximum range 7.60 km (4.72 miles).
Charge five: velocity 391 mps (1,283 fps), maximum range 9.15 km (5.68 miles).
Charge six: velocity 470 mps (1,542 fps), maximum range 10.67 km (6.63 miles).

Firing smoke shell weighing 14 kg (30 lb 13 oz):
Charge six: velocity 460 mps (1,509 fps), maximum range 10.60 km (6.58 miles).

Firing coloured-smoke shell weighing 14.75 kg (32 lb 8 oz):
Charge six: velocity 470 mps (1,542 fps), maximum range 10.40 km (6.46 miles).

Firing hollow charge shells, models 'A', 'B', or 'C':
Charge six: velocity 496 mps (1,627 fps), engagement range 1.50 km (0.93 miles).

The Propelling Bag Charge

The propellant charge bags for the 10.5-cm le FH 18 gun came in six sizes, numbered one to six. Charge bag six contained the most explosives and was used for long-range targets. Charges one to five were contained in silk bags. Once the range to target was known, the correct charge bag was placed in a short (normally brass) cartridge case and a cap was placed on the top. Charge six bag was supplied separately and replaced portions one to five when necessary (the others were emptied out of the case and the charge six bag placed inside).

The charge bags were filled with different amounts of *Nitroglyzerin Blättchen Pulver* (nitroglycerin-flaked powder) explosives. This was abbreviated to 'Ngl.Bl.P.'. Sometimes, they were filled with a different type of explosives called *Diglykolnitrat Blättchen Pulver* (Diglycolnitrate flaked powder). This was abbreviated to 'Digl.Bl.P.'.

The composition of the charges bags for the 10.5-cm leFH 18 gun was as follows:

Charge one: 180 gm (6.35 oz) Ngl.Bl.P., or 241 gm (8.50 oz) Digl.Bl.P.
Charge two: 55 gm (1.94 oz) Ngl.Bl.P., or 59 gm (2.08 oz) Digl.Bl.P.
Charge three: 60 gm (2.12 oz) Ngl.Bl.P., or 70 gm (2.47 oz) Digl.Bl.P.
Charge four: 115 gm (4.06 oz) Ngl.Bl.P., or 118 gm (4.16 oz) Digl.Bl.P.
Charge five: 185 gm (6.52 oz) Ngl.Bl.P., or 192 gm (6.77 oz) Digl.Bl.P.
Charge six: 784 gm (1 lb 12 oz) Ngl.Bl.P., plus 200 gm (7.05 oz) Digl.Bl.P. as an igniter.

The 10.5-cm *leichte Feldhaubitze* 18 *Mündungbremse* (10.5-cm leFH 18M)

The Soviet 7.62-mm M1939 and the Allied 25-pounder artillery guns had longer ranges than the standard German 10.5-cm leFH 18 gun. Its maximum range was 10.67 km (6.63 miles). The British 25-pounder had a maximum range of 12.25 km (7.61 miles). The Soviet 7.62-mm had a maximum range of 13.25 km (8.23 miles). German gun crews could be shelled by the enemy but had no way of retaliating as they did not have the range.

The 10.5-cm leFH 18M was fitted with a '*Mündungbremse*' muzzle brake to allow longer range charges to be fired and reduce the amount of recoil on the gun by dispersing the gases in the barrel sideways in addition to out of the front of the barrel each time a shell was fired. This helped increase the operational life of the gun barrel. Some surviving museum examples have had the muzzle brake removed.

The recoil buffer valves were changed on this newer version of the gun and the air pressure in the recuperator was increased. The length of the gun was now 3.31 m (10 feet 2 inches). It now had a velocity of 540 mps (1,772 fps). The larger propellant bag charge increased the gun's maximum range from 10.67 km (6.63 miles) to 12.32 km (7.65 miles).

The 10.5-cm *leichte Feldhaubitze* 18/40 (10.5-cm leFH 18/40)

In March 1942, the German Army demanded that a lighter version of the 10.5-cm leFH 18 light field howitzer be manufactured to make transportation easier. The gun required the same performance specifications.

This 10.5-cm *leichte Feldhaubitze* 18 *Mundungbremse* (10.5-cm leFH 18M) is on display at the Museo Militaria, The Artillery, Engineer and Signals Museum, Hämeenlinna, Finland. Notice it is fitted with a muzzle brake.

The weight reduction was achieved by changing the gun carriage and making adjustments to the barrel and top recoil system. The lighter gun carriage used on the 7.5-cm PaK 40 anti-tank gun was modified and fitted to the artillery gun.

The '*Mündungbremse*' muzzle brake was improved by welding projecting flanges to the original 18M version. This enabled a greater amount of the explosive gases to be deflected to the side and thus reducing the stresses on the gun barrel. Some surviving museum examples have had the muzzle brake removed.

This version of the gun was mounted on the 10.5-cm leFH 18/40/2 (Sf) *auf Geschützwagen* III/IV self-propelled artillery gun.

Specifications

Calibre: 105 mm (4.13 inches).
Length of gun: 2.94 m (9 feet 8 inches).
Length of bore: 2.71 m (8 feet 11 inches).
Rifling: Thirty-two grooves, right-hand increasing twist, 1/18 to 1/12.
Breech mechanism: horizontal sliding block, percussion fired.
Traverse: 56 degrees.
Elevation: -6 to +40 degrees.
Weight in action: 1.95 tonnes (1.92 tons).

The 15-cm *schwere Feldhaubitze* 13 (15-cm sFH 13)

The German arms manufacturer Krupp started to design this gun in 1913 as a replacement for the 15-cm sFH 02. It could fire shells over a longer distance. It had a bigger gun and it was fitted with a gun shield to protect the crew. It had a long single box trail with a foldable spade at the end, which was forced into the ground to help the gun handle recoil. It was mounted on the 15-cm sFH 13 (Sf) *auf Geschützwagen* Lorraine Schlepper(f) artillery SPG.

This gun was a heavy field howitzer used by the German and Ottoman Empires during the First World War. They were not fitted with a '*Mündungbremse*' muzzle brake.

It had a horizontal sliding breech block and a hydro-spring variable recoil mechanism. It used two-part ammunition. The shell was loaded first and then the cased charge rammed into the breech after it. The gun crew had a choice of seven charge bags to put into the cased charge cartridge. Charge bag seven was the most powerful: it was loaded with the most propellant explosive and was used for long-range targets. It had a muzzle velocity of 381 mps (1,250 fps) and a maximum firing range of 8.60 km (5.34 miles) when the charge seven bag was used.

A number of these horse-drawn towed artillery guns survived into the Second World War as reserve and training weapons. Some of these guns were in use with the Belgium and Dutch Armies. When the Germans occupied both countries in May 1940, they were seized and taken back into service with the German Wehrmacht. They were not used for front-line service during the Second World War except the ninety-four guns that were mounted on the

This 15-cm *schwere Feldhaubitze* 13 (15-cm sFH 13) is on display at the Fort Sill, Lawton, Oklahoma, USA. (*Jon Bernstein*)

15-cm sFH 13 (Sf) *auf Geschützwagen* Lorraine Schlepper(f) self-propelled artillery gun.

When the Second World War started, the only data on this weapon available to Allied troops were based on a British GHQ Intelligence summary written in August 1918.

It required a team of six horses to tow one gun. Three men were needed to look after them. The gun crew normally consisted of six additional men. They would ride on the gun limber.

Specifications

Calibre: 150 mm (5.92 inches).
Length of gun: 2.54 m (8 feet 4 inches).
Length of bore: 2.09 mm (6 feet 10 inches).
Rifling: Thirty-two grooves, right-hand increasing twist, 1/45 to 1/22.4.
Breech mechanism: horizontal sliding block, percussion fired.
Traverse: 9 degrees.
Elevation: -4 to +45 degrees.
Weight in action: 2.25 tonnes (2 tons).

The 15-cm *schwere Feldhaubitze* 18 (15-cm sFH 18)

The 15-cm sFH 18 heavy field howitzer was designed to replace the First World War 15-cm sFH 13. Two German arms manufacturers submitted designs for the new gun. The German High Command instructed Krupp to build the gun barrel and Rheinmetall-Borsig AG to manufacture the gun carriage.

The weight of the gun was twice as heavy as the 15-cm sFH 13. The gun barrel length was longer: now 4.39 m (14 feet 5 inches). It had double the muzzle velocity of its predecessor and had a maximum firing range of 13.3 km (8.26 miles). The 15-cm sFH 13 only had a maximum range of 8.5 km (5.28 miles).

It was so heavy that the gun had to be towed in two parts by two teams of eight horses: one would pull the gun carriage and the other the gun barrel. When the gun was complete and in a firing position, it weighed 5.5 tonnes (approximately the weight of five Ford Fiesta cars).

When motorised tracked prime mover artillery tractors became available the gun was not broken down into two parts. They were towed on four wheels.

It used two-part ammunition. The shell went into the barrel first followed by the charge canister. Each 15-cm sFH 18 crew were issued with a range of eight different charge bags, two more than the gun crews of the 10.5-cm leFH 18 howitzers.

This 15-cm *schwere Feldhaubitze* 18 (15-cm sFH 18) is on display at the Museo Militaria, The Artillery, Engineer and Signals Museum, Hämeenlinna, Finland.

This 15-cm *schwere Feldhaubitze* 18 (15-cm sFH 18) heavy field howitzer is on display at the Norfolk Tank Museum, Forncett St Peter, Norwich, England. Notice it is fitted with a muzzle brake. This is an addition following the Second World War.

Specifications

Calibre: 150 mm (5.91 inches).
Length of gun: 4.49 m (14 feet 9 inches).
Length of bore: 4.12 m (13 feet 6 inches).
Rifling: Forty grooves, right-hand increasing twist, 1/36 to 1/18.
Breech mechanism: horizontal sliding block, percussion fired.
Traverse: 64 degrees.
Elevation: -3 to +45 degrees.
Weight in action: 5.51 tonnes (5.43 tons).

Maximum Firing Range

Firing standard high-explosive shell weighing 43.50 kg (95 lb 14 oz):
Charge one: velocity 210 mps (689 fps), maximum range 4.00 km (2.48 miles).
Charge two: velocity 230 mps (755 fps), maximum range 4.70 km (2.92 miles).
Charge three: velocity 250 mps (820 fps), maximum range 5.52 km (3.43 miles).
Charge four: velocity 278 mps (912 fps), maximum range 6.62 km (4.11 miles).
Charge five: velocity 320 mps (1050 fps), maximum range 8.20 km (5.09 miles).
Charge six: velocity 375 mps (1230 fps), maximum range 9.72 km (6.04 miles).
Charge seven: velocity 440 mps (1444 fps), maximum range 11.40 km (7.08 miles).
Charge eight: velocity 495 mps (1624 fps), maximum range 13.25 km (8.23 miles).

Propelling Charges

The explosives in charge bags one to six were put in an artificial silk bag, contained in the cartridge case. Charge bags seven and eight were delivered separately in a metal canister. Charge bag seven came in a tubular silk bag that was 21.5 cm (8.46 inches) long, and 15 cm (5.92 inches) in diameter. Charge bag eight was supplied in a flat circular bag about 17.5 cm (6.89 inches) in diameter.

Charge bags one to six were produced by combining bags. When a target was 6.62 km (4.11 miles) away, the gun needed a charge four bag. The charge seven bag replaced all the previous increments in the cartridge case, and for charge eight, the last bag was added.

The charge igniter used Ngl.Bl.P.; some 550 gm (19.40 oz) of Ngl.Bl.P was used with all charge bags. The charge bags were filled with different amounts of Digl.Bl.P.

The composition of the charges bags for the 15-cm leFH 18 gun was as follows:

Charge one: 62 gm (2.19 oz) Digl.Bl.P.
Charge two: 122 gm (4.30 oz) Digl.Bl.P.
Charge three: 124 gm (4.37 oz) Digl.Bl.P.
Charge four: 208 gm (7.34 oz) Digl.Bl.P.
Charge five: 312 gm (11.00 oz) Digl.Bl.P.
Charge six: 475 gm (16.75 oz) Digl.Bl.P.
Charge seven: 2.39 kg (84.30 oz) Digl.R.P.
Charge eight: 768 gm (27.08 oz) Digl.R.P.

This 15-cm *schwere Feldhaubitze* 18 (15-cm sFH 18) heavy field howitzer is on display at Tim Issac's Cobbaton Combat Collection, North Devon, England. It is fitted with a post-Second World War muzzle brake.

The 15-cm *schwere Feldhaubitze* 18/1 (15-cm sFH 18/1)

When the 15-cm sFH 18 gun mount was adapted to be fitted into the 15-cm sFH 18/1 (Sf) *auf Geschützwagen* III/IV Hummel (Sd.Kfz.165) self-propelled artillery gun, a number '1' was added after the number '18' to differentiate it from the towed howitzer.

The 15-cm *schweres Infanteriegeschütz* 33 (15-cm sIG 33)

This 15-cm short barrelled heavy infantry gun was a cut-down version of the larger 15-cm sFH 18. The lighter weight made it easier to transport. It was fitted to the 15-cm sIG 33 (Sf) *auf Fahrgestell Panzerkampfwagen* II, 15-cm sIG 33 L/11 (Sf) *auf Fahrgestell Panzerkampfwagen* III Ausf. H, 15-cm sIG 33 (Sf) *auf Geschützwagen* 38(t) *Grille* (Sd. Kfz.138/1), and 15-cm sIG 33/2 (Sf) *auf Jagdpanzer* 38(t) self-propelled artillery guns.

The German words '*schweres Infanteriegeschütz*' translates to 'heavy infantry gun'. This was abbreviated to sIG. The number '33' is confusing as the gun was first issued in 1927, not 1933. The gun was designed by the German arms manufacturer Rheinmetall-Borsig AG. It was developed following a German Army directive of 1920 stating the need for a 7.5-cm and 15-cm howitzer.

It had a standard horizontal sliding-block breech mechanism. To help deal with the recoil once the gun was fired, it had a hydropneumatic recoil system in a cradle below the barrel. The pivot point of the gun was to the rear. This enabled a high degree of elevation.

This 15-cm *schweres Infanteriegeschütz* 33 (15-cm sIG 33) heavy infantry gun is on display at the Imperial War Museum Duxford, England.

The gun carriage was a two-wheeled split trail type: it had two legs that opened at the rear to give greater stability when the gun fired. They were locked together when the gun was towed. Two spring balancing presses were fitted to the outside of the trails to counteract the weight of the muzzle. The gun carriage wheels were made of pressed metal with solid rubber tyres.

Although a lot smaller than the 15-cm sFH 18 heavy field howitzer, it was still too large for the infantry to handle with ease. To try and make it more mobile, lighter alloy wheels replaced the original heavier metal wheels. This practice was stopped when the Second World War broke out as light alloys were in high demand for the manufacture of aircraft for the Luftwaffe. The heavier wheels were reissued.

The 15-cm sIG was a robust and reliable artillery infantry support gun that could deliver a powerful high-explosive fragmentation shell at distances up to 4.7 km (2.92 miles). It remained in service until the end of the war. A number of different European nations still used this weapon after the war.

Specifications

Calibre: 150 mm (5.91 inches).
Length of gun: 1.74 m (5 feet 8 inches).
Length of bore: 1.65 m (5 feet 4 inches).
Rifling: Forty-four grooves, right-hand uniform twist, 1/21 Breech mechanism: horizontal sliding block, percussion fired.
Traverse: 11.5 degrees.
Elevation: 0–73 degrees (some sources give 4–75 degrees).
Weight in action: 1.70 tonnes (1.67 tons) (standard type); 1.55 tonnes (1.52 tons) (lightweight type).

Maximum Firing Range

Firing standard high-explosive shell weighing 38 kg (83 lb 12 oz):
Charge one: velocity 122 mps (400 fps), maximum range 1.47 km (0.91 miles)
Charge two: velocity 152 mps (500 fps), maximum range 2.12 km (1.32 miles).
Charge three: velocity 186 mps (610 fps), maximum range 3 km (1.86 miles).
Charge four: velocity 210 mps (690 fps), maximum range 3.75 km (2.33 miles).
Charge five: velocity 230 mps (755 fps), maximum range 4.37 km (2.71).
Charge six: velocity 240 mps (787 fps), maximum range 4.70 km (2.92 miles).

Propelling Charge

The six silk cloth charge bags were filled with different amounts of Ngl.Bl.P. explosives. They were filled with Digl.Bl.P., which was used as the igniter.

Bag number six would have the most amount of propellant explosive and could fire the high-explosive shell the longest distance. If the target was around 3.75 km away, the crew would need to use a charge four bag. They could put in a single charge four bag into the cartridge case or two charge two bags or one charge one bag followed by a charge three bag. A cap would be placed on the end of the cartridge case. The HE shell would be loaded first followed by the cartridge case.

The composition of the charges bags for the 15-cm sIG 33 gun was as follows:

Charge one: 40 gm (1.41 oz) Ngl.Bl.P., plus 42 gm (1.48 oz) Digl.Bl.P.
Charge two: 85 gm (3.00 oz) Ngl.Bl.P., plus 42 gm (1.48 oz) Digl.Bl.P.
Charge three: 101 gm (3.56 oz) Ngl.Bl.P., plus 42 gm (1.48 oz) Digl.Bl.P.
Charge four: 103 gm (3.63 oz) Ngl.Bl.P., plus 42 gm (1.48 oz) Digl.Bl.P.
Charge five: 117 gm (4.13 oz) Ngl.Bl.P., plus 42 gm (1.48 oz) Digl.Bl.P.
Charge six: 130 gm (4.58 oz) Ngl.Bl.P., plus 42 gm (1.48 oz) Digl.Bl.P.

7.5-cm *Gebirgshaubitze* 36 (7.5-cm GebH 36)

Mountain troops needed a small, light, and easily transportable howitzer. German weapons manufacturer Rheinmetall-Borsig AG came up with a new design to replace the First World War mountain divisions' (*Gebirgs Divisionen*) guns like the Skoda 7.5-cm *Gebirgskanone* model 15 (7.5-cm GebK 15). It was called the 7.5-cm *Gebirgsgeschütz* 36 (7.5-cm GebG 36) light mountain howitzer. Between 1938 and 1945, documents show that 1,193 were built. It became the standard light gun of the German mountain divisions, of both German Army and Waffen-SS during the Second World War. Experiments were conducted to mount it to the back of a *Geschützwagen Raupenschlepper Ost* (RSO/3) tracked prime mover and use it as a self-propelled artillery gun.

It was a standard German horizontal sliding breech block, percussion-fired gun. It was fitted with a muzzle brake to disperse the explosive gases in the gun barrel more efficiently and help reduce recoil. It used a variable recoil system that shortened the recoil as the elevation increased to stop the gun breach hitting the ground.

Rear trunnions were added to lengthen the distance between the breech and the ground. The recoil mechanism was hydropneumatic, with both buffer and recuperator positioned below the barrel in a cradle. It had split trail legs with foldable spades at the end to make the gun more stable when firing. When it was being towed, the trail legs were moved and locked together. The spades at the end of the split trail legs were removable.

To keep the weight down, the gun carriage was fitted with light-alloy disc wheels with rubber rims. Early versions of the gun had wooden-spoked wheels. No protective gun shield was fitted; this was a weight-saving decision. It weighed 750 kg (1,654 lb) so was within the weight limit for loads capable of being carried by vehicles like the RSO.

Although it was a ground design, the 7.5-cm GebK 15 suffered from a lack of stability at low angles of elevation because of the lightness of the gun. Charge five could only be fired when the angle of elevation was in excess of 15 degrees; otherwise, the gun would jump dangerously.

The 7.5-cm *Gebirgsgeschütz* 36 (7.5-cm GebG 36) mountain gun (*Rade Nagraisalović*)

The strength of the recoil would force the gun's trail spades to act as a fulcrum and lever the wheels upwards. When the gun was fired at higher angles, it performed better as the ground absorbed any residual recoil forces not absorbed by the recoil system.

It used two-part ammunition and with four bag charges of propellant that were added together depending on the range of the target. A larger, fifth charge bag was used on its own when the target was at the limit of the howitzer's maximum range. It fired a high-explosive shell that had a maximum range of 9.15 km (5.68 miles). The shell weighed 5.75 kg (12.68 lb). A good gun crew were able to produce a rate of fire of six to eight rounds per minute.

It could also fire smoke shells. In an emergency, a hollow charge armour-piercing anti-tank round could be fired at a tank. It weighed 5.83 kilograms (12.9 lb) with a maximum range of 9.25 km (5.74 miles) but it was only effective at short range.

This mountain gun could be broken down into six separate parts each having a maximum weight of 300 pounds. This ability enabled the weapon to be easily transported by pack animal or in an aeroplane. Tests were carried out to see if it could be carried in one piece by the *Raupenschlepper Ost* (RSO/3).

The gun's 56-inch barrel was of a monobloc construction. To enable larger and more powerful charges to be used to increase the range of the gun without damaging the gun barrel, it was fitted with a perforated, six-baffled muzzle brake.

German *Gebirgs-Artillerie* Regiments (mountain artillery regiments) were organised into four gun batteries. A battalion would consist of two or three batteries.

Specifications

Calibre: 75 mm (2.95 inch)
Length of barrel: 1.45 m (4 feet 9 inches)
Length of bore: 97.2 cm (3 feet 2 inches)
Rifling: Twenty-eight grooves, uniform right-hand twist
Carriage: Split trail
Weight: 750 kg (1,654 lb) shipped in eight parts
Traverse: 40 degrees
Elevation: -2 to 70 degrees
Muzzle velocity: 475 mps (1,560 fps)
Range: 9.25 km
Crew: Five
Rate of fire: Six–eight rounds per minute
Produced: 1938–44, by Rheinmetall-Borsig AG at R. Wolf of Magdeburg-Bückau
Number built: Over 1,193

Propelling Charge

The five silk cloth charge bags were filled with different amounts of Digl.R.P (*Diglykol Röhrenpulver*). The abbreviation 'R.P.' refers to the shape of the propellant—in this case, small hollow tubes. *Diglykol* is a cool-burning, less corrosive alternate to nitroglycerin.

Bag number five would have the most amount of propellant explosive and could fire the HE shells the longest distance. The HE shell would be loaded first followed by the cartridge case.

The composition of the charges bags for the 7.5-cm GebH 36 gun was as follows: there were five charges, of which one to four were issued in the cartridge case; charge five was issued separately. On the occasions that charge five was used, the charge bag was inserted in the case in place of the discarded one to four.

Charge one: 123 gm (4.34 oz) Digl.R.P.
Charge two: 158 gm (5.57 oz) Digl.R.P.
Charge three: 200 gm (7.05 oz) Digl.R.P.
Charge four: 275 gm (9.70 oz) Digl.R.P.
Charge five: 530 gm (18.69 oz) Digl.R.P. plus 10 gm (0.35 oz) NZ Man P.

10.5-cm *Gebirgshaubitze* 40 (10.5-cm GebH 40)

The Austrian weapons manufacturer Böhler submitted designs for a heavy mountain gun in response to a German Army directive. They were in competition with the German company Rheinmetall. They won the contract.

As with most mountain guns, it could be broken down into smaller parts to help with transportation over difficult terrain. It could also be towed as a complete weapon if the

This 10.5-cm *Gebirgshaubitze* 40 (10.5-cm GebH 40) mountain gun is on display at the Central Museum of the Great Patriotic War 1941–1945, Park Pobedy, Moscow.

roads were good. The gun appears to be too big for the gun carriage but that was a deliberate weight saving aspect of the design. No gun shield was fitted. This was another weight-saving design feature. The wheels were of light alloy with solid tyres.

When broken down, it could be transported in five separate pieces and put on the backs of mules or horses. Special single-axle trailers were developed to be towed by small vehicles like the NSU Kettenrad tracked motorcycle. The gun could be broken down and towed on four of these trailers. Tests were carried out to see if it could be carried in one piece by the *Raupenschlepper Ost* (RSO/1).

To give stability, the trail legs could be split when the gun was fired. Removable spades could be attached to the end of the trails and dug into the ground to help cope with the gun recoil.

The 10.5-cm GebH 40 was fitted with a '*Mündungbremse*' muzzle brake to allow longer range charges to be fired; it also reduced the amount of recoil on the gun by dispersing the gases in the barrel sideways in addition to out of the front of the barrel each time a shell was fired. This helped increase the operational life of the gun barrel.

The Böhler 10.5-cm *Gebirgshaubitze* 40 (10.5-cm GebH 40) was issued to mountain regiments in 1942. After the Second World War, a number of European Armed forces still deployed this weapon. They remained in service until the middle of the 1960s.

Specifications

Calibre: 105 mm (4.13 inches).
Length of gun: 3.44 m (11 feet 9 inches).

Length of bore: 2.87 m (9 feet 4 inches).
Rifling: Thirty-two grooves, increasing right-hand twist, 1/31.5 to 1/21.9.
Breech mechanism: horizontal sliding block, percussion fired.
Traverse: 51 degrees.
Elevation: -5.5 to +71 degrees.
Weight in action: 1.66 tonnes (1.63 tons).

Maximum Firing Range

Firing standard high-explosive fragmentation shell FH Gr Al, weighing 14.50 kg (31 lb 15 oz):
Charge one: velocity 210 mps (689 fps), maximum range 3.84 km (2.38 miles).
Charge two: velocity 225 mps (738 fps), maximum range 4.47 km (2.77 miles).
Charge three: velocity 255 mps (837 fps), maximum range 5.30 km (3.29 miles).
Charge four: velocity 285 mps (935 fps), maximum range 6.54 km (4.06 miles).
Charge five: velocity 340 mps (1,116 fps), maximum range 8.00 km (4.97 miles).
Charge six: velocity 425 mps (1,394 fps), maximum range 9.67 km (6 miles).
Charge seven: velocity 565 mps (1,854 fps), maximum range 16.74 km (10.40 miles).

Firing hollow charge shell 10-cm Gr 39:
Charge six: velocity 460 mps (1,509 fps), maximum range 8 km (4.97 miles).

Firing star shell 10.5-cm *Leuchtgeschoss*:
Charge six: velocity 480 mps (1,575 fps), maximum range 6.90 km (4.28 miles).
This gave an ejection height of 500 m (1,640 feet) for optimum illumination.

Propelling Charge

The six silk charge bags were filled with different amounts *Nitrozellulose-Manöverpulver* (nitrocellulose propellant with the addition of potassium nitrate). This was abbreviated to 'NZ Man P'. On charge bags three to six, *Gudol Blättchenpulver* was also added. *Diglykol*, with the addition of nitroguanidine, lowers the temperature at which the gunpowder ignites, reducing barrel wear, as well as acting as a stabiliser. *Blättchenpulver* means flakes (literally leaf powder). This was abbreviated to 'Gudol Bl P'.

Bag number seven would have the most amount of propellant explosive and could fire the HE shell the longest distance. The HE shell would be loaded first followed by the cartridge case.

The composition of the charges bags for the 10.5-cm GebH 40 gun was as follows:

Charge one: 20 gm (0.71 oz) NZ Man P, 230 gm (8.11 oz). Digl.Bl.P., 10 gm (0.35 oz) flash reducer.

Charge two: 20 gm (0.71 oz) NZ Man P, 267 gm (9.42 oz). Digl.Bl.P., 10 gm (0.35 oz) flash reducer.

Charge three: 20 gm (0.71 oz) NZ Man P, 267 gm (9.42 oz). Digl.Bl.P., 55 gm (1.94 oz) Gudol Bl P, 15 gm (0.53 oz) flash reducer.

Charge four: 20 gm (0.71 oz) NZ Man P, 267 gm (9.42 oz). Digl.Bl.P., 155 gm (5.47 oz) Gudol Bl P, 20 gm (0.71 oz) flash reducer.

Charge five: 20 gm (0.71 oz) NZ Man P, 267 gm (9.42 oz). Digl.Bl.P., 310 gm (10.93 oz) Gudol Bl P, 20 gm (0.71 oz) flash reducer.

Charge six: 20 gm (0.71 oz) NZ Man P, 267 gm (9.42 oz). Digl.Bl.P., 610 gm (21.51 oz) Gudol Bl P, 25 gm (0.88 oz) flash reducer.

Charge seven: 20 gm (0.71 oz) NZ Man P, 2.06 kg (4 lb 8 oz) Digl.Bl.P., no flash reducer.

21-cm *Mörser* 18

The First World War German Army 21-cm Lange *Mörser* 16 was a heavy howitzer capable of firing a high-explosive fragmentation shell that weighed 113 kg (249 lb) on to a target at a maximum range of 11.1 km (6.89 miles). It stayed in service until 1940. The German weapons manufacturer Krupp submitted a design for a 21-cm Mörser howitzer to replace it. The gun's maximum range was increased to 16.7 km (10.37 miles). This was a drastic improvement. It was also more accurate. Krupp was given the contract to manufacture this new weapon. They were first issued to German artillery units in 1939. A prototype self-propelled artillery gun chassis and superstructure were built to carry this gun. The chassis was built using Tiger tank parts.

The problem with large older artillery pieces was lack of repeated accuracy once the target had been located and hit. The enormous recoil exerted upon the gun mount or carriage often shifted the gun's position after each shell was fired.

German 21-cm *Mörser* 18 on display at the US Army Field Artillery Museum, Fort Sill, Lawton, Oklahoma, USA. (*Jon Bernstein*)

The Krupp design team came up with a radical solution for this problem. It is known as the 'dual recoil' system. The gun barrel moved backwards and forwards along the barrel cradle in the normal way controlled by a conventional hydropneumatic recuperator system to handle the initial recoil. What made this design new was the way the gun mounting slid along the carriage on rails. This meant that both systems together absorbed nearly 100 per cent of the recoil force.

The 21-cm *Mörser* 18 heavy artillery mortar canon was mounted on a very effective steady gun platform that was ideal to maintain accurate heavy round high-explosive bombardment of a target at long range.

The carriage carried an integral circular firing platform. It was lowered to the ground when the gun was being ready for deployment by the crew. Three caster wheels were then cranked down on to the platform. They took the weight of the gun. The front gun carriage wheels were now off the ground. This allowed the gun to be traversed left and right.

The box trail end of the gun was fitted with a large spade. This was lifted off the ground when the gun needed to be turned in a different direction or adjusted slightly. The rear caster wheel jack was used to raise the spade and rear section of the gun trails off the floor.

Krupp also designed the 17-cm *Kanone* 18 heavy artillery mortar cannon to go on the same gun carriage. It had a better maximum range than the 21-cm *Mörser* 18. It could fire a lighter 68-kg (149-lb) high-explosive fragmentation shell on to a target 29.6 km (18.4 miles) away. Due to the higher velocity of the gun, the smaller shell had nearly the same explosive power as the larger 113-kg shell used by the 21-cm *Mörser* 18. It was ideal for counter battery bombardment because of its long range.

The German Army artillery regiments and generals made its views known to the Army High Command. They wanted more 17-cm *Kanone* 18 guns rather than the 21-cm *Mörser* 18. In 1942, production was cancelled in favour of the 17-cm *Kanone* 18.

For long-distance transportation, the 21-cm *Mörser* 18 could be split into two loads. The gun barrel was disconnected from the gun carriage and carried by a separate transport vehicle. The carriage was equipped with a series of ramps and winches designed to help remove the barrel. For short distances, the gun could be towed in one piece. It had its faults. It took too long to pack away to be ready for an advance and to set up in a new battery position. This meant that the guns might not be available when the troops at the front needed artillery support. As it took a long time to pack away, during a retreat, the guns were in danger of being overrun by enemy tanks and troops.

It could fire three different shells. This gun was not designed as an anti-tank weapon. It used separate-loading cased charges. The shell was loaded followed by a cased cartridge full of the correct amount of explosive propellant for the range of the target. It only fired high-explosive fragmentation shells. The 21-cm Rö Gr 42 Be high-explosive projectile weighed 192 kg (423 lb 5 oz). It was designed as a bunker buster. It was known as a Röchling anti-concrete projectile. In tests, it could penetrate 4 m (13 feet 1 inch) of concrete but on the battlefield, it was found that better results were obtained, possibly because of the variable quality of the concrete it was firing at. It was a very long projectile that had folding fins at the rear. The outer metal case broke away from the shell after firing: it had a three-part multi-grooved sabot at the shoulder of the shell that fell away after it left the muzzle of the gun barrel. The shell was 2.59 m (8 feet 6 inches) long and had an explosive head of 2.85 kg (6 lb 4 oz).

The 21-cm Gr 37 high-explosive shell was also developed as a bunker buster concrete penetrating shell. It weighed 120.3 kg (265 lb 3 oz). The 21-cm Gr Be high-explosive shell was also designed to be used against reinforced strong points. It weighed 121.4 kg (267 lb 10 oz). The 21-cm Gr. 18 Stg shell was lighter. It weighed 113 kg (249 lb 2 oz). It was not specifically designed to penetrate concrete. It was a standard high-explosive shell for use against enemy 'soft-skinned' targets like lorries, half-tracks, artillery batteries, machine gun nests, mortar units, and infantry.

Specifications

Calibre: 211 mm (8 inches).
Length of gun: 6.51 m (21 feet 4 inches).
Length of bore: 6.07 m (26 feet 6 inches).
Rifling: Sixty-four grooves, right-hand increasing twist, 1/35 to 1/30.
Breech mechanism: horizontal sliding block, percussion fired.
Traverse: 16 degrees on the carriage, 360 degrees on the circular firing platform.
Elevation: 0 to 70 degrees.
Weight in action: 16.70 tonnes (16.43 tons).

Maximum Firing Range

Firing standard 21-cm Gr 18 high-explosive shell weighing 113 kg (249 lb 2 oz):
Charge one: velocity 225 mps (739 fps), maximum range 4.65 km (2.88 miles)
Charge two: velocity 255 mps (837 fps), maximum range 5.85 km (3.63 miles)
Charge three: velocity 290 mps (951 fps), maximum range 7.32 km (4.54 miles)
Charge four: velocity 355 mps (1165 fps), maximum range 9.80 km (6.08 miles)
Charge five: velocity 440 mps (1,444 fps), maximum range 12.50 km (7.76 miles)
Charge six: velocity 565 mps (1,854 fps), maximum range 16.70 km (10.37 miles)

Firing 21-cm Gr Be high-explosive shell weighing 121.4 kg (267 lb 10 oz):
Charge one: velocity 218 mps (715 fps), maximum range 4.35 km (2.70 miles)
Charge two: velocity 247 mps (810 fps), maximum range 5.50 km (3.41 miles)
Charge three: velocity 281 mps (922 fps), maximum range 7 km (4.34 miles).
Charge four: velocity 344 mps (1,129 fps), maximum range 9.60 km (5.96 miles).
Charge five: velocity 426 mps (1,398 fps), maximum range 12.40 km (7.7 miles)
Charge six: velocity 550 mps (1,805 fps), maximum range 16.72 km (10.38 miles)

Propelling Charge

The six charge bags were filled with different amounts *Diglykol Ringpulver* explosives. This was abbreviated to Digl.Rg.P. *Diglykol* is a smokeless propellant, based on diethylene glycol dinitrate and nitrocellulose, stabilised with centralite. Potassium sulphate was sometimes added to reduce flash. The word *Ringpulver* refers to the shape of the propellant: perforated disks (literally ring powder). Charge one was a double bag charge of different explosives for ballistic reasons.

Charge one: 1.00 kg (2 lb 3oz) Digl.Bl.P. plus 1.37 kg (3 lb) Digl.Rg.P.
Charge two: 0.527 kg (1 lb 3 oz) Digl.Rg.P.
Charge three: 0.697 kg (1 lb 9 oz) Digl.Rg.P.
Charge four: 1.480 kg (3 lb 4 oz) Digl.Rg.P.
Charge five: 2.124 kg (4 lb 11 oz) Digl.Rg.P.
Charge six: 15.537 kg (34 lb 4 oz) Digl.R.P.

The very long 21-cm Rö Gr 42 Be high-explosive Röchling projectile had a special charge. It was put in a single bag inside a cased cartridge. It was made up of 13.04 kg (28 lb 12 oz) of Digl.R.P. explosives. The term *Röhrenpulver* means hollow tube. These charge bags were marked in red to make them distinctive.

17-cm *Kanone* 18 in *Mörserlafette*

The German Army 17-cm K18 heavy artillery mortar canon was produced between 1941 and 1945. Over 300 were manufactured. They were used by heavy artillery batteries on many fronts to support infantry and panzer divisions.

During the First World War, the German artillery used a 21-cm *Mörser* 16 mortar. It was replaced in 1939 by the 21-cm *Mörser* 18. It could fire a 113-kg (249-lb) high-explosive fragmentation shell on enemy targets that were 16.7 km (10.37 miles) away. A prototype self-propelled artillery gun chassis and superstructure were built to carry this gun. The chassis was built using Tiger tank parts.

The German Army wanted a similar heavy artillery piece that had a longer maximum range. The German weapons manufacturer Krupp's engineers and design team came up with a smaller 17.3-cm calibre gun but with an increased velocity. It used the same gun carriage as the 21-cm Mörser 18. It was given the name 17-cm *Kanone* 18 in *Mörserlafette*. This was shortened to 17-cm K18 in MrsLaf. The German word '*Mörserlafette*' translates to mortar gun carriage.

It could fire a 68-kg (149-lb) high-explosive fragmentation shell on to a target 29.6 km (18.4 miles) away. Due to the higher velocity of the gun, the smaller shell had nearly the same explosive power as the larger 113 kg shell of the 21-cm *Mörser* 18. Another advantage of the 17-cm *Kanone* 18 was that supplying ammunition would be easier. The shells were smaller so more could be fitted into the ammunition supply vehicles. It used two-part ammunition: separate-loading cased charges. The shell was put in the breach first followed by the cartridge with the correct charge bag full of propellant for the range of the target.

The inspection team of German artillery officer of the Wa Prüf 4 (German Army development and testing department, artillery branch) of the *Heereswaffenamt* (HWA),

17-cm *Kanone* 18 in *Mörserlafette* on display at the US Army Field Artillery Museum, Fort Sill, Lawton, Oklahoma, USA. (*Jon Bernstein*)

German High Command's centre for technical weapons development was impressed with the results and Krupp were given the contract to start production.

The 17-cm *Kanone* 18 used the same 'dual recoil' system as previously mentioned in the section on the 21-cm *Mörser* 16. This meant it was more accurate than earlier guns as nearly all of the gun's recoil was absorbed.

It could fire three different shells. This gun was not designed as an anti-tank weapon but if it was surprised by enemy tanks, it could fire a 17-cm Pzgr 73 armour-piercing shell that weighed 71 kg (156 lb 8 oz). It only had a range of around 1 km (0.62 miles) but could penetrate 255 mm of 30-degree sloped armour.

The standard HE fragmentation shell was called the 17-cm K Gr 39. It weighed 68 kg (149 lb 14 oz). It only had a maximum range of 28 km (17.39 miles). For longer range targets, the 17-cm K Gr 38 Hb shell was loaded. It was slightly lighter at 62.8 kg (138 lb 7 oz) and when fired using charge bag number four, it had a maximum range of 29.60 km (18.39 miles).

For long-distance transportation, the 17-cm *Kanone* 18 could be split into two loads. The gun barrel was disconnected from the gun carriage and carried by a separate transport vehicle. The carriage was equipped with a series of ramps and winches designed to help remove the barrel. For short distances, the gun could be towed in one piece. Although this gun was very effective, it had a major flaw. It took too long to set up into a new battery position and to pack up for further transportation. Guns were lost and crews became casualties as positions were overrun by enemy forces during a counterattack.

Specifications

Calibre: 173 mm (6.81 inches).
Length of gun: 8.53 m (28 feet).
Length of bore: 8.10 m (26 feet 6 inches).
Rifling: Forty-eight grooves, right-hand increasing twist, 1/50 to 1/30.
Breech mechanism: horizontal sliding block, percussion fired.
Traverse: 16 degrees on the carriage, 360 degrees on the circular firing platform.
Elevation: 0 to 50 degrees.
Weight in action: 17.52 tonnes (17.24 tons).

Maximum Firing Range

Firing long-range high-explosive shell weighing 62.80 kg (138.47 lb):
Charge one: velocity 620 mps (2,034 fps), maximum range 18.30 km (11.37 miles)
Charge two: velocity 740 mps (2,428 fps), maximum range 22.70 km (14.10 miles)
Charge three: velocity 860 mps (2,822 fps), maximum range 28 km (17.39 miles)
Charge four: velocity 925 mps (3,035 fps), maximum range 29.60 km (18.39 miles)

Propelling Charge

The five cloth charge bags were filled with different amounts of Digl.R.P. The four levels of charges were made up from a combination of five bags. Each was given a different name depending on the amount of explosive they had in them: *hauptkart* (main bag), *sonderkart* (special bag), and *vorkart* (front bag):

Vorkart 4: 1.35 kg (2 lb 15 oz) Digl.R.P.
Sonderkart 2: 4.89 kg (10 lb 12 oz) Digl.R.P.
Vorkart 3: 12.95 kg (28 lb 9 oz) Digl.R.P.
Sonderkart 1: 14.98 kg (33 lb) Digl.R.P.
Hauptkart: 16.00 kg (35 lb 4 oz) Digl.R.P.

The four charges were made from various combinations of five bags being placed in the propellant cartridge case, charge four being the most powerful:

Charge one: *Sonderkart* 1 bag placed in the cartridge case.
Charge two: *Sonderkart* 2 bag placed in the cartridge case.
Charge three: *Hauptkart* bag placed in the cartridge case, with *Vorkart* 3 bag loaded into the gun chamber ahead of it.
Charge four: the *Hauptkart*, *Vorkart* 3 and *Vorkart* 4 bag (pushed into an internal cavity in the centre of the *Vorkart* 3 bag)

2

15-cm sIG 33 L/12 (Sf) *auf Geschützwagen* I Ausf. B 'Bison' (Sd.Kfz.101)

The German Army's experiences in the Spanish Civil War and the invasion of Poland on 1 September 1939 made it realise that highly mobile armoured divisions needed the support of artillery guns that could keep pace with the tanks over rugged, sometimes wet and muddy, ground.

The early Panzer IV tanks were armed with a short barrelled 7.5-cm artillery howitzer. Their job was to provide close artillery support for the 3.7-cm anti-tank gun-armed Panzer III tanks. Some situations needed a more powerful artillery gun.

The infantry of the *Panzergrenadiers* also needed armoured close-support howitzers to deal with enemy vehicles and artillery batteries, strong points, machine gun nests, fortified farmhouses, and other buildings. Entries in post-battle reports after the campaign in Poland drew attention to the lack of heavy artillery support able to cope with the fast pace of the panzer divisions.

The German arms manufacturer Alkett started to work on a design project to meet this need. In 1940, Panzer I light tanks were classed as being not fit for front-line service as a battle tank. Alkett decided to keep costs down by reusing the Panzer I tank chassis and mount an artillery howitzer on the top.

A Panzer I Ausf. B light tank chassis was used for the prototype. The design concept called for the gun to be quickly removed from the tank chassis and worked as a stand-alone gun. This would enable the gun still to be used if the tank chassis became immobile through mechanical breakdown or battle damage.

In this configuration, it acted as a *Waffenträger* (weapon carrier). It became more famous for being deployed as a *Selbstfahrlafette Geschützwagen* (literally translated as self-propelled carriage gun vehicle). The letters 'Sf' stand for *Selbstfahrlafette*.

The Panzer I tank chassis was not very long. Alkett's designers were limited as to the artillery guns that would fit on top of the tank. After some trials, their final choice was the standard 15-cm *schweres Infanteriegeschütz* 33 (15-cm sIG 33). This 15-cm short-barrelled gun was a cut-down and lighter version of the larger 15-cm sFH 18.

The gun mounted the chassis via ramps at the rear. The gun carriage wheels were kept on. They were not removed like in some of the First World War gun carriage designs. The wheels were locked down to the top of the chassis once they had been moved into two individual metal rectangular troughs.

This gun only had one rear trail. It did not have two trail legs that split. The end of the gun trail rested on top of the rear section of the engine compartment. Keeping the engine at the back of the chassis in its original configuration and keeping the wheels on the gun meant that this self-propelled artillery gun had a very high profile. This made it an easy target to spot on an open farmland battlefield.

The gun crew were protected from small arms fire and shrapnel from high-explosive shells by a 13-mm (0.51-inch) bolted, armoured-plate, open-top and rear fighting compartment. This vehicle was not designed to be used on the front line when fighting enemy tanks or in urban situations. The crew were not protected in these scenarios.

The superstructure covered the driver's position on the sides and over the top. This meant that the driver had limited vision. He could only see forward. A member of the gun crew would have to walk in front of the SPG and direct the driver when reversing. The driver would need to rely on crew members to tell him if there was traffic on the road to his left or right when crossing a road junction.

The gun crews could fire at targets they could see but normally, they would be called to provide covering fire on enemy targets some miles away over the horizon. They would fire by plotting map grid references provided by forward artillery observers, tank, or infantry commanders.

Although wartime footage shows this artillery vehicle being used as an assault gun, firing at enemy strongholds in a bomb-damaged city, it was not designed for this role. The crew would be at risk from enemy snipers and hand grenades. *Sturmpanzers* (armoured assault guns), like the 15-cm StuIG 33B and the StuG III SPGs, were more suitable for urban assault as they had fully enclosed armoured fighting compartments.

Gun laying was made by the means of a Rblf36 sight. The gun had a 5.5-degree traverse left and 5.5-degree right. It could be depressed down 4 degrees and elevated 75 degrees. Once the gun was loaded, wicker baskets were strapped to the top of the track guard at the rear. These contained the three 15-cm high-explosive shells and three cased cartridge powder charges. There was only room for three rounds. Additional ammunition had to be towed in a trailer or carried in by another vehicle. When the battery was set up, the gun crew would go to the ammunition carrying vehicle and stack up on the floor a supply of shells and cased cartridge powder charges behind the SPG, ready for orders to open fire. The end of the gun barrel was covered in a cap when not in use to keep out mud and grit.

It was powered by a Maybach NL 38 TR six-cylinder water-cooled 2.8-litre 100-hp petrol engine. It used a ZF F.G. 31 transmission. It had five forward gears and one reverse. The suspension system used coil/leaf springs. It had a maximum road speed of 40 km/h (25 mph). Its operational range was 170 km (105 miles) on road and around 115 km (71 miles) across the country.

After the prototype passed trials, Alkett was instructed to commence production. Thirty-eight were manufactured. They were available for issuing to German Army units in February 1940.

The Name

The official designation was 15-cm sIG 33 L/12 (Sf) *auf Geschützwagen* I Ausf. B (Sd.Kfz.101). It was also known as 15-cm *schweres Infanteriegeschütz* (mot S) *auf Panzerkampfwagen* I (Sd.Kfz.101) *ohne Aufbau Ausführung* B. After the war, it was given the much shorter nickname the 'Bison'. It was not known by that name during the Second World War.

The name Bison was used by model kit companies. It is believed they were responsible for giving this SPG that name to help sell their product. It is sometimes wrongly called the '*Sturmpanzer* I'. As previously mentioned, this self-propelled artillery gun is not an assault tank.

Operational History

They were first deployed during the French campaign, serving in the *schwere Infanteriegeschütz Kompanies* (heavy infantry gun companies: sIG Kp) 701, 702, 703, 705, and 706. They were independent units attached to panzer divisions. Each company had three platoons. Each platoon consisted of two 15-cm sIG 33 L/12 (Sf) *auf Geschützwagen* I Ausf. B self-propelled artillery guns and a Sd.Kfz. 10 (*Sonderkraftfahrzeug*: special motorised vehicle) half-track prime mover. Not all of the gun crew could travel on the SPG over long distances as it was too small. They were transported around the battlefield and on long road journeys in the half-track. This vehicle was also used to carry ammunition and could also tow a trailer loaded with shells.

During the Battle of France in May 1940, they were assigned to different panzer divisions. Each one had six 15-cm sIG 33 L/12 (Sf) *auf Geschützwagen* I Ausf. B self-propelled artillery guns. They were formed on 30 March 1940 and were ready for service on 22 April 1940.

701.s.I.G.Kp was part of 9. *Panzerdivision.*
702.s.I.G.Kp was part of 1. *Panzerdivision.*
703.s.I.G.Kp was part of 2. *Panzerdivision.*
704.s.I.G.Kp was part of 5. *Panzerdivision.*
705.s.I.G.Kp was part of 7. *Panzerdivision.*
706.s.I.G.Kp was part of 10. *Panzerdivision.*

All of the SPGs survived the invasion of France campaign. Three heavy infantry gun companies later saw action during the campaign of the Balkans, in Yugoslavia, and then in Greece.

On 22 June 1941, Germany invaded the Soviet Union. This attack was called Operation Barbarossa. They had earlier been redeployed to Romania, to be ready in time for the invasion.

Few survived Operation Barbarossa. Only the 701st is recorded as participating in the following year's Operation *Fall Blau* (Case Blue) in June 1942. It was a two-pronged attack against the oil fields of Baku as well as an advance in the direction of Stalingrad along the Volga River, to cover the flanks of the advance towards Baku.

On 30 June 1943, the last two 15-cm sIG 33 L/12 (Sf) *auf Geschützwagen* I Ausf. B self-propelled artillery guns were reported as still being fit for service in the depleted sIG Kp.704, attached to the 5. *Panzerdivision*.

Conclusion

They suffered from a lot of mechanical breakdowns due to the weight of the gun and its recoil overloading of the chassis, engine, and suspension. Its tall superstructure gave it a very high profile. It was hard to hide and it often became a target for enemy guns. The lack of all-round armoured protection for the gun crew resulted in casualties. Due to the limited room for ammunition stowage, they were dependent on their ammunition supply vehicles. On the bad roads of Russia and Ukraine, they had problems keeping up with the fully tracked vehicle. Quite often the guns could not be used because of lack of shells.

Surviving SPGs

There are no surviving examples of the 15-cm sIG 33 L/12 (Sf) *auf Geschützwagen* I Ausf. B 'Bison' (Sd.Kfz.101)

Specifications

Length: 4.42 m (14 feet 6 inches)
Width: 2.17 m (7 feet 1 inch)
Height: 2.70 m (8 feet 10 inches)
Total weight, battle ready: 7 tonnes (6.88 tons)
Crew: Four (driver, commander, gunner, and loader)
Propulsion: Maybach 6-cyl NL 38 TR water-cooled, 3.8-litre petrol, 100-hp engine
Armament: 15-cm sIG 33 L/11 howitzer (5.9 inches),
Armor: From 4 to 13 mm
Speed (on/off road): 40/15 km/h (25/9 mph)
Range; 170 km (105 miles)
Total production: thirty-eight

15-cm sIG 33 (Sf) *auf Geschützwagen* I *Ausf* B (Bison) Sd.Kfz. 101 of unit 704, attached to the 5th Panzer Division in Grey livery. (*David Bocquelet*)

15-cm sIG 33 (Sf) *auf Geschützwagen* I Ausf B (Bison) Sd.Kfz. 101 of unit 704, attached to the 5th Panzer Division in winter whitewash livery. (*David Bocquelet*)

15-cm sIG 33 (Sf) *auf Geschützwagen* I *Ausf* B (Bison) Sd.Kfz. 101 of unit 704, attached to the 5th Panzer Division with a camouflage livery. (*David Bocquelet*)

3

10.5-cm leFH 16 (Sf) *auf Geschützwagen* Mk VI(e)

With the Fall of France in May 1940, after the German *Blitzkrieg* invasion, all British Army Expeditionary Force (BEF) tanks and vehicles had to be left behind. At Dunkirk, there was only room on the ships for soldiers to escape back to England. When the vehicles were abandoned, the British troops tried to make them unusable so the German Army could not turn them against their previous owners.

Not all their efforts were successful. Many captured tanks were in a repairable condition. Parts were taken from different vehicles to make one vehicle serviceable. The German mechanics managed to repair a number of British Vickers Mk VIb and Mk VIc light tanks.

The German Army needed artillery that could keep up with the tanks of the panzer division. A decision was made to use some of the captured tanks, including the Vickers Mk VI light tanks, as self-propelled artillery guns.

These were known as *Beutepanzers* (trophy tanks) and given the official designation of *Leichte Panzerkampfwagen* Mk VIB 735(e) or *Leichte Panzerkampfwagen* Mk VIC 736(e). They were used for combat, reconnaissance, internal police security, and tank crew training. The letter 'e' in brackets referred to the country of origin of the captured tank, in this case, England.

Design and Production

A German engineer and artillery officer called *Oberleutnant* Alfred Becker, who had been attached to the tank producing factory of Alkett in Berlin, organised the mounting of six First World War-era 10.5-cm leFH 16 light field howitzers on the top of captured Vickers Mk VI tanks.

The gun crew were protected by an armoured casement that surrounded the 10.5-cm leFH 16 howitzer. Its armour thickness ranged from 11–22 mm (0.43–0.89 inches). That was enough to protect the crew from small arms fire and shrapnel fragments from high-explosive shells and mortar rounds.

This self-propelled gun was not designed to fight on the front line. It was a self-propelled artillery gun that would stay behind the main assault and fire high-explosive shells over the heads of the troops and tanks at targets given to it as grid references on a map by forward observers.

It was not an anti-tank gun, though it did carry a few armour-piercing rounds for self-defence at close quarters. A machine gun could be fitted to the front of the superstructure to the right of the gun. The crew had their own weapons that they could fire over the top of the fighting compartment if enemy troops got too close.

The fighting compartment was open-topped; it did not have an armoured roof. This kept the weight down and gave the commander all-round vision. The crew would cover the top with a tarpaulin in bad weather and also in very hot weather, to provide shade.

To help deal with recoil from the gun when it was fired, the gun crew deployed a metal support frame attached to the rear of the vehicle. It was a square-shaped metal frame that was strengthened by two crossbars and at the bottom had V-shaped 'spades' at the bottom that would dig into the earth.

The 'spades' only needed to be used to help stabilise the gun when using propellant charge bags four and five for longer range targets. Closer targets, which required less powerful charge bags one to three, could be engaged without them folded down. The recoil of the gun did not affect the vehicle's position despite its low weight.

The commander sat at the rear of the vehicle, on the left side, behind the gunner. He had access to a range finding periscope mounted to the side of the vehicle. The gunner's gun sight poked out above the top of the forward gun shield and armour casement. The loader sat on the right side of the vehicle.

Although only six were built, it is reported that an artillery command vehicle version was built on a captured Vickers Mk IV light tank chassis. It had an enclosed superstructure with a cupola on the top. As yet, no drawings or photographs have been found of this vehicle.

Active Service

The six converted Vickers Mk IV light tanks, now fitted with the long-range 10.5-cm leFH 16 artillery howitzer, were placed in the 12th Battery of the 227th Artillery Regiment, 227th Infantry Division. They were divided up into two platoons of three. The battery was also equipped with tracked ammunition carriers. The commanding officer was the same German engineer, Alfred Becker, now promoted to captain, who had been involved in their construction.

The German documents refer to this unit as an assault battery though it was not used in that role. The armour on the front of these vehicles was not strong enough to take a direct hit from an enemy tank or anti-tank gun. These SPGs were used as artillery.

The 227th Infantry Division performed coastal defence and internal occupation security duties after the Fall of France in Normandy, near the port of Le Havre, from July 1940 until late summer 1941. The gun crews had time to train on their new vehicle before being posted to the Army Group North, 1st Army Corps on the Eastern Front and taking part in the heavy fighting in and around Leningrad (St Petersburg), Russia.

The 227th Infantry Division was transferred to the German Army Group North, arriving in the autumn of 1941. They relieved the 39th Motorised Corps who had been in position in the forests south of Lake Ladoga.

At this part of the battle, the German commanders were expecting a Soviet 54th Army counterattack to try and end the siege surrounding Leningrad. The 10.5-cm LeFH 16 *auf Geschützwagen* Mk VI(e) artillery SPGs of the 12th Battery took up defensive positions where they could fire on the routes it was believed the Red Army troops and tanks would take. They had time to zero in their guns, so as soon as the advancing Soviet forces were spotted by forward observers, they could cover the area in high-explosive shells.

They were temporarily transferred to the nearby 254th Infantry Division and provided mobile artillery support when they took part in the 54th Army's offensive of 20 October 1941. The 10.5-cm LeFH 16 *auf* GW Mk VI(e) were used for numerous artillery fire missions between 23 and 24 October 1941, firing over 200 rounds. It appears that the 15th Battery was divided into three platoons of two guns each.

The unit suffered its first casualties. Four men including Captain Becker were wounded when the SPGs were used for direct fire infantry support. This was a role not suitable for this artillery weapon due to the lack of heavy protective armour on these vehicles.

The two 10.5-cm LeFH 16 *auf Geschützwagen* Mk VI(e) artillery SPGs of the 1st Platoon were very active in late October 1941. They were ordered to support the 11th Infantry Division as they attacked towards the villages of Volkhov and Pogostye. Soviet infantry surrounded the SPGs and the tank crews had to use their personal weapons and hand grenades to fight off the enemy troops.

On 11 November 1941, near the village of Khotovskaya Gorka, Soviet light tanks engaged the 1st Platoon Artillery SPGs. A German battle report confirmed the Soviet report that one of the vehicles was hit sixteen times, but its armour was never penetrated. The tanks they confronted were Soviet T-40 light tanks of the 2nd Tank Brigade. Luckily for the German gun crews, the T-40 was only armed with machine guns.

Three men were killed on 15 November 1941 when the battery was ordered to act as assault guns while supporting an unsuccessful attack of the 223rd Infantry Division. One of the vehicles had to be left in no man's land after being heavily damaged by a mine. It was recovered three days later.

The 12th Battery had fired over 1,300 shells, which is just over 200 shells per gun crew, between November and mid-December 1941.

The 12th Battery conducted more artillery support fire missions during the winter and spring of 1942 as the 227th Infantry Division continued to fight around the village of Pogostye.

Three of the six 10.5-cm LeFH 16 *auf Geschützwagen* Mk VI(e) SPGs were knocked out on 16 February 1942, when KV-1 tanks of the 124th Tank Battalion of the Soviet 54th Army attacked German units. The 10.5-cm armour-piercing ammunition given to the German gun crews failed to knock out the Red Army heavily armoured KV-1 tanks.

The three remaining vehicles provided close support as troops moved through forest roads near Pogostye in March 1942. They used their HE shells to destroy a machine gun nest and fire on a Soviet infantry column they encountered in the woods.

One of the remaining three vehicles was reported not fit for active service and could not be repaired after the close quarters fighting during March. The 12th Battery now only had two vehicles. They were used to break through an enemy encirclement. One was knocked out by a Soviet anti-tank rifle.

The last remaining 10.5-cm LeFH 16 *auf* GW Mk VI(e) self-propelled gun was destroyed by a Red Army tank of the 98th Tank Brigade as it tried to provide protective covering fire for one of the German supply routes.

A German Army post-battle report dated November 1941 covered the role played by the 12th Battery, 227th Artillery assault gun battery in the battle of Leningrad (St Petersburg) in October 1941 near the village of Mga south of the city.

The reporting officer was impressed with the high-explosive shell performance of the 10.5-cm LeFH 16 howitzer over the 7.5-cm HE shells. It was found that the two-part ammunition used on the larger more modern guns, such as the 10.5-cm LeFH 18, could be used up to canister charge five. This helped in dealing with the logistics of ammunition supplies.

The overall evaluation of the 10.5-cm LeFH 16 *auf* GW Mk VI(e) in combat was positive. It was a good stable firing platform and it had good cross-country performance. The small dimensions of the vehicle and low weight allowed it to cross weak bridges. The engine was reliable and suffered relatively few mechanical breakdowns. It had an economical fuel consumption figure of 4 miles per gallon. This was better than most other armoured fighting vehicles of this time. Fuel was in short supply on the Eastern Front so this feature of the vehicle was welcomed. Due to these findings, more self-propelled artillery guns were built although on different larger tank chassis. *Oberleutnant* Becker, later Major Becker, would play a prominent role in the development on many of these SPGs, after he recovered from his wounds.

Battery Markings

On some photographs taken of 10.5-cm LeFH 16 *auf* GW Mk VI(e) on the Eastern Front, there are numbers and letters on the side of the vehicle. The letters 'Gp' are an abbreviation for the word '*Geschützpanzer*', which translated as gun tank or self-propelled gun. The number is the vehicle identification number given to the vehicle by the Regiment: Gp 4 would be gun tank number four in an artillery battery of six vehicles.

Surviving SPGs

There are no surviving examples of a 10.5-cm leFH 16 (Sf) *auf Geschützwagen* Mk VI(e) self-propelled artillery gun.

The 10.5-cm LeFH 16 *auf Geschützwagen* Mk VI(e) self-propelled artillery gun built on a Vickers Mk VI light tank chassis. (*David Bocquelet*)

10.5-cm LeFH 16 *auf Geschützwagen* Mk VI(e) *Beutepanzer* SPG conversion of a captured British Vickers Mk VI light tank. Gp1 indicates that this vehicle was the first gun tank in a battery of six vehicles. (*David Bocquelet*)

10.5-cm LeFH 16 *auf Geschützwagen* Mk VI(e) in winter whitewash livery. (*David Bocquelet*)

Specifications

Length: 4 m (13 feet 1 inch)
Width: 2.20 m (7 feet 3 inch)
Height: 2 m (6 feet 7 inch)
Total weight, battle ready: 6.5 tonnes (6.39 tons)
Crew: Four (commander, driver, gunner, and loader)
Propulsion: Meadows six-cylinder gasoline/petrol engine, 88 hp
Top road speed: 35 mph (56 km/h)
Off road speed: 25 mph (40 km/h)
Operational range (road): 174 miles (280 km)
Main Armament: 10.5-cm leFH 16 howitzer
Armor: 4–14 mm
Total production: Six

4

7.5-cm *Gebirgshaubitze* 36 *auf Geschützwagen Raupenschlepper Ost* (RSO/3)

The Germans experimented with transporting and mounting a number of different guns on the back of the *Raupenschlepper Ost* light 'prime mover' tracked vehicle. The name *Raupenschlepper Ost* translates as 'caterpillar tractor east'. It is commonly abbreviated to just RSO.

The prototypes were shown to the Army. The *Raupenschlepper Ost* 7.5-cm PaK 40 tank destroyer self-propelled gun went into production. Between eighty and ninety were produced. Most saw action on the Eastern Front. A version of the RSO that carried a 2-cm Flak 38 anti-aircraft gun mounted to the floor of the rear wooden cargo bay also saw service.

At present no documentation has been found relating to the mounting and carrying of artillery guns on the back of the *Raupenschlepper Ost* even though there are surviving photographs of four different prototypes: the 7.5-cm GebH 36 *auf* RSO/3; the 7.5-cm GebH 34 *auf* RSG; the 10.5-cm GebH 40 *auf* RSO/1; and the 15-cm sIG 33 *auf* RSO/3. It is not clear if these prototypes were going to be used as a *Waffenträger* (weapon carrier) or as a self-propelled gun.

If they were used as a *Waffenträger*, then how was the gun dismounted? There is photographic evidence that the guns were loaded on the vehicle by a winch attached to a freestanding metal frame on a hard surface. Another photograph shows a *Raupenschlepper Ost* reversed back towards an earthen ramp so the gun could be pushed on to the back of the vehicle.

On a battlefield, it would be difficult to build a ramp quickly or make sure there was a hard surface for a winch and frame to be constructed on to enable the guns to be unloaded. The guns were heavy and if the load bearing frame was put together on soft earth its legs would sink into the ground under the weight.

If these prototype vehicles were intended to be used as a *Selbstfahrlafette Geschützwagen* (self-propelled gun), the problem the engineers would have to overcome was the recoil. With the artillery gun mounted in the back of the vehicle, they were very top heavy and had a high centre of gravity. There was a danger that the RSO would topple over.

It can be theorised that two of the prototypes were intended to be used as artillery SPGs but tests showed the RSO chassis was not strong enough to take the gun recoil so they were never put into production. This is supported by the fact that on the photographs of the 7.5-cm GebG 36 *auf* RSO/03, the side panels are down and it can be seen that the gun wheels had been clamped to the deck of the vehicle plus the gun 'tails' had been shortened. The 7.5-cm *Gebirgshaubitze* 34 *auf Gebirgsraupenschlepper* (RSG) also carried a similarly sized howitzer.

The other two prototypes seen in photographs are carrying much bigger 10.5-cm and 15-cm howitzers. There is no evidence that these guns were bolted to the wooden cargo bay of the RSO tracked vehicle so that it could be fired. The gun's split trail legs had not been modified to fit the length of the vehicle. They protrude out the back and the rear 'spades' are carried in the back of the vehicle for use when the gun is set up on land again. The RSO tracked vehicle is being used as a *Waffenträger* weapon carrier in these examples.

The *Raupenschlepper Ost* RSO tracked vehicle

The RSO light 'prime mover' tracked vehicle had a very basic suspension design with all steel wheels and just four small leaf springs. This made it cheap and easy to produce. It had high ground clearance and excellent performance in poor terrain. It was a tracked version of the Steyr 1.5-tonne truck. It could carry a 1,500-kg (3,307-lb) load in its cargo bay.

The Steyr-Daimler-Puch manufacturing company designed the *Raupenschlepper Ost* (RSO) to be used to tow field guns and transport supplies over rough ground in muddy waterlogged and snowy conditions. They were in production between October 1943 and May 1944: Steyr-Daimler-Puch produced 2,600 vehicles; Klöckner-Deutz-Magirus (KHD) manufactured 12,500; Auto-Union made a further 5,600; and Gräf and Stift constructed 4,500 RSOs. They were used extensively on the Eastern Front.

There were four main variants. The RSO/01, RSO/02, and RSO/PaK 40 were powered by a 3.5-l Steyr V8 gasoline/petrol 70-hp engine. The RSO/03 had a better performing Deutz F4L514 5.3-l four-cylinder air-cooled diesel engine although produced lower horsepower at 66 hp.

The RSO/1 had a fully enclosed pressed steel rounded cab with a wooden rear cargo bay. The RSO/2 had a flat sided metal cab. The RSO/3 was manufactured by KHD at their Magirus factory and had a simplified slab-sided metal cab. The RSO/PaK 40 had a lightly armoured low-profile steel cab to enable the 7.5-cm PaK 40 anti-tank gun mounted on the rear flat rear bed wooden cargo bay to fire forward.

7.5-cm *Gebirgshaubitze* 36 *auf Raupenschlepper Ost* (RSO/3)

To mount the 7.5-cm *Gebirgsgeschütz* 36 (7.5-cm GebG 36) light mountain howitzer on the back of the *Raupenschlepper Ost* tracked vehicle cargo bay the spades at the end of the split trail legs were removed. The legs were also cut down in length to allow the back-tail gate to be raised. The wheels were bolted to the wooden floor in a special semicircular frame. This gun was meant to be fired from the back of the RSO. It could no longer be dismounted and fired from the ground without having new split trail legs fitted. It could not function as a *Waffenträger* weapon carrier. It was a self-propelled gun prototype.

The gun was built by Rheinmetall to replace the First World War mountain divisions' (*Gebirgs Divisionen*) guns. Between 1938 and 1945, records show 1,193 were built. It was a standard German horizontal sliding breech block gun with a muzzle brake. It used a variable recoil system that shortened the recoil as the elevation increased to stop the gun breach hitting the ground. Rear trunnions were added to lengthen the distance between the breech and the ground. The recoil mechanism was hydropneumatic, with both buffer and recuperator positioned below the barrel.

To keep the weight down, the gun was fitted with light-alloy disc wheels with rubber rims. No protective gun shield was fitted to save weight. It weighed 750 kg (1,650 lb) so it was within the cargo weight limit of the RSO.

When used on the ground, the 7.5-cm GebG 36 would jump when fired at low angles, because of its lightness. The strength of the recoil would force the gun's trail spades to act as a fulcrum and lever the wheels upwards. The shell canister bag charge five, the largest propellant increment, was forbidden to be used at near horizontal angles under 15 degrees because the gun would jump excessively. When the gun was fired at higher angles, it performed better as the ground absorbed any residual recoil forces not absorbed by the recoil system. On the back of the RSO, the vehicle's suspension, tracks, and the ground had to absorb the force of the recoil from the gun.

The 7.5-cm *Gebirgsgeschütz* 36 mountain howitzer used two-part ammunition, with four bag charges of propellant that were added together depending on the range of the target. A larger fifth charge bag was used on its own when the target was at the limit of the howitzers maximum range. It fired a high-explosive 5.83-kg (12.9-lb) shell that had a maximum range of 9.25 km (10,120 yards). It could also fire smoke shells and in an emergency a hollow charge armour-piercing rounds at short range. A good gun crew was able to produce a rate of fire of six to eight rounds per minute.

This mountain gun could be broken down into six separate parts, each having a maximum weight of 300 pounds. This ability enabled the weapon to be easily transported by pack animals or in an aeroplane.

The gun's 56-inch barrel was of a monobloc construction. To enable larger and more powerful charges to be used and to increase the range of the gun without damaging the gun barrel, it was fitted with a perforated, six-baffled muzzle brake.

7.5-cm GebH *36 auf Gebirgsraupenschlepper* (RSG)

When the Steyr-made *Gebirgsraupenschlepper* (RSG) mountain troop-tracked vehicle is parked next to the larger *Raupenschlepper Ost* (RSO/3) vehicle, it is very noticeable that the RSG is a lot smaller. There is a photograph showing a 7.5-cm *Gebirgshaubitze* (GebH) mountain howitzer mounted on the back of the RSG. Only one photograph has been found so far of this prototype self-propelled artillery gun.

The problem is that the caption that went along with this photo identified the gun on the back as a captured Belgium Army Swedish-built Bofors 75-mm Model 1934 mountain gun (Canon de 75 mle 1934). It was recorded as a 7.5-cm *Gebirgshaubitze* 34 *auf* RSG, but this gun was not fitted with a circular perforated muzzle brake.

It can be theorised that the howitzer on the back is the same gun used on the 7.5-cm *Gebirgshaubitze* 36 *auf Raupenschlepper Ost*, which does have a circular perforated muzzle brake. Just like on the other vehicle, it would have had its split tail legs cut to fit the length of the wooden cargo bay and the wheels clamped to the floor so that the gun could be fired from the back of the vehicle.

10.5-cm *Gebirgshaubitze* 40 *auf Raupenschlepper Ost* (RSO/1)

There is a poor-quality photograph showing a 10.5-cm *Gebirgshaubitze* 40 (10.5-cm GebH 40) mountain howitzer on the back of a *Raupenschlepper Ost* (RSO/1).

In the picture, it looks like the vehicle has been backed up to a ramp of earth. There appears to be wooden planks spanning the gap between the top of the earth mound and the back of the RSO/1. Its tail gate is hinged down and so are the wooden side panels. These wooden planks would have been used to enable the gun to be pushed on to the back of the vehicle.

Unlike on the photographs of the 7.5-cm *Gebirgshaubitze* 36 *auf Raupenschlepper Ost* (RSO/3), there is no visible evidence that the larger 10.5-cm GebH 40 gun had been fixed to the wooden floor of the cargo bay. The split trail legs had not been cut and shortened. They projected over the rear of the vehicle.

There was no special semicircular locking wheel frame in use. The spades that were normally fitted to the end of the split trail legs had not been attached to. Their triangular shape can be seen at the back of the gun.

Was this photograph taken of an early live firing trial to see if the RSO/1 could take the gun recoil or just to see if it could take the weight of the gun? It is not known as no documents have been found so far.

In the other surviving photographs, the gun is seen on the back of the RSO/1 with the wooden side panels in the up position, the split trail legs sticking out the back, and the tail spades loaded on the rear with the tail gate panel in the down position.

The RSO/1 tracked vehicle has the manufacturing company's name and logo on the side. This is a factory vehicle, not one that has been sold to the Army. It is safe to assume that it is the company, Steyr-Daimler-Puch, who was experimenting with showing that the 10.5-cm GebH 40 mountain howitzer can be transported on the back of their vehicle.

In three better-quality photographs, it looks like a frame and winch had been used to lift the gun on to the back of the RSO/1. These photographs would suggest that this vehicle was being used as a *Waffenträger* weapon carrier. At present, there is no evidence the vehicle was used as a *Selbstfahrlafette Geschützwagen* (self-propelled artillery gun), and fired the gun from the back of the cargo bay, as there are no visible mountings or fixings to secure the gun to the vehicle.

There only appears to be photographs of a 10.5-cm *Gebirgshaubitze* 40 mountain howitzer on the back of one RSO tracked vehicle. It is highly unlikely that the experiment was a success as the weight of the gun exceeded the designed load weight of the vehicle. The gun weighed 1,660 kg (3,660 lb) and the RSO's load weight limit was 1.500 kg (3,307 lb). The RSO's centre of gravity would have been significantly raised. Both these things would have made the vehicle tricky to drive.

15-cm sIG 33 *auf Raupenschlepper Ost* (RSO/3)

There is only one photograph currently available of a 15-cm sIG 33 (*schweres Infanterie Geschütz* 33), the standard German heavy infantry gun in the Second World War, loaded on the back of a *Raupenschlepper Ost* (RSO/3) tracked vehicle. The split trail legs can be seen sticking out the back of the vehicle. There had been no attempt to cut them to fit the length of the RSO/3's wooden cargo bay.

This was not a test to see if the 15-cm sIG 33 howitzer could be fired from the back of the vehicle. The gun was too big and the RSO/3 would not have been able to handle the violent recoil. This vehicle was not a German *Selbstfahrlafette Geschützwagen*, a self-propelled artillery gun. It was almost assuredly a trial to see if the gun could be carried on the back of the RSO/3.

The experiment most likely failed, as the weight of the gun exceeded the designed load weight of the vehicle. The gun weighed 1,800 kg (4,000 lb) and the RSO's load weight limit was 1,500 kg (3,307 lb). The RSO's centre of gravity would have been significantly raised. Both these things would have made the vehicle sluggish and difficult to manoeuvre. The RSO/3 was not a suitable vehicle to be a *Waffenträger* weapon carrier for the 15-cm sIG 33 howitzer.

Conclusion

The most plausible theory is that the Steyr-Daimler-Puch manufacturing company wanted to win a lucrative German government contract to build self-propelled artillery guns using their cheap to produce *Raupenschlepper Ost* light tracked vehicle and RSG. They exhibited four prototype vehicles that had different artillery howitzers mounted on the back to the government inspectors.

Two of the guns used were too big for the RSO tractor. The 7.5-cm mountain howitzer was light enough and could be mounted to the floor of the wooden cargo bay at the rear of the RSO and RSG vehicles. These prototypes seemed viable as artillery SPGs.

At the time, there was competition from other vehicle and arms manufacturers who wanted to be awarded a government contract to build self-propelled artillery guns. Their designs used sturdier German tank chassis or captured enemy armoured fighting vehicles on which to mount artillery guns. Steyr-Daimler-Puch's designs were not approved for production.

Surviving RSOs

There are no surviving examples of a 7.5-cm *Gebirgshaubitze* 36 *auf Geschützwagen Raupenschlepper Ost* (RSO/3). There are lots of surviving RSO-tracked supply vehicles around the world in different museums and in private collections.

The original *Raupenschlepper Ost* tracked supply vehicle. (*David Bocquelet*)

15-cm sIG 33 *auf Raupenschlepper Ost* (RSO/3). (*David Bocquelet*)

Specifications RSO

Length: 7.19 m (14 feet 6 inches)
Width: 3 m (6 feet 6 inches)
Height: 2.87 m (8 feet 6 inches)
Total weight unladen: 7.73 tonnes (7.61 tons)
Armament: 7.5-cm *Gebirgsgeschütz* 36
As Weapon Carrier: 15-cm sIG33
Armour: none
Track width: 33 cm/61 cm with snow plates (13 inches/24 inches)
RSO/1-2 Propulsion: 3.5-l Steyr V8 gasoline/petrol 70-hp engine
RSO/3 Propulsion: Deutz F4L514 5.3-l four-cylinder air-cooled 66-hp diesel engine
Fording depth: 34 inches
Top road speed: 30 km/h (18 mph)
Operational range (road): 300 km (155 miles)

5

10.5-cm leFH 18/1 (Sf) *auf Geschützwagen* IVb (Sd.Kfz.165/1)

Design and Production

The large German engineering company Krupp had been involved in designing and manufacturing ammunition, artillery, armoured fighting vehicles, and infantry weapons for the German Army for many years. They had received their first order to build 135 Panzer I tanks in 1933.

In 1939, they turned their attention to the problem of how to mount a large artillery gun on a tank chassis. Their solution was to fix a 10.5-cm LeFH 18/1 artillery light field howitzer on top of a shortened Panzer IV tank chassis in an open-topped turret. It was given the official designation 10.5-cm leFH 18/1 (Sf) *auf Geschützwagen* IVb. The chassis had only had six road wheels rather than the normal eight. The six wheels were 570 mm in diameter instead of the normal 470-mm-diameter road wheels. It also only had three (not four) top track rollers. Krupp presented their design to the military on 14 September 1939.

The turret could not fully traverse 360 degrees. It could only traverse 70 degrees: 35 degrees to the left and 35 degrees to the right. It was not a tank, even though it may resemble one at first glance.

Two test vehicles were built. They were given the internal factory designations of V1 and V2 (nothing to do with the later flying bomb and rocket). The letter 'V' was an abbreviation for the German word '*Versuchs*', which translates as 'trial' or 'prototype'. The German Army liked what they saw and put an order in for ten more to be built in the autumn of 1941. The first one was completed at the Krupp Grusonwerk factory in August 1942, three more in September, four in October, one in November, and one in December 1942. The ten vehicles were accepted into the Army in January 1943.

The first two test vehicles V1 and V2 were powered by a Maybach HL66P engine that produced 188 hp. The next ten SPGs were given a more powerful engine: a Maybach HL90 P20k twelve-cylinder engine that produced 320 hp. Their chassis numbers ranged from 150631 to 150640. They were given the official designation of 10.5-cm leFH 18/1 (Sf) *auf Geschützwagen* IVb (Sd.Kfz.165/1). This was shortened to 10.5-cm leFH 18/1 (Sf) *auf* GW IVb or Pz.Sf.IVb.

The turret armour was not thick. It ranged from 14.5–30 mm (0.57–1.18 inches). It provided the gun crew protection from small arms fire, high-explosive shell fragments, and mortars. The open top reduced weight and allowed the commander all-round vision. In bad weather, a tarpaulin was fixed over the top of the open turret. It was also used in very hot weather. When not in use, it was rolled up on top of the rear turret storage box.

The Panzer IV hull-mounted machine gun was removed to free up more storage space. This vehicle was not designed to be an assault gun or anti-tank gun. It was to provide mobile artillery support that could keep up with the attacking panzer divisions.

It was not envisaged to be fighting on the front line. It could fire high-explosive shells over long distances on to enemy targets it was given via a grid reference by forward observation units. If Soviet infantry got too close they could use their personal weapons or retreat as fast as they could to a safer location.

The gun was issued with a few armour-piercing rounds for self-defence if they were surprised by Soviet tanks. They only worked at close range and were ineffective against the front armour of the more heavily armoured T-34 and KV-1 tanks.

The commander sat at the rear of the turret on the left side behind the gunner. He had access to a range finding periscope mounted to the side of the vehicle. The gunner's gun sight poked out above the top of the forward gun shield and armour casement. The loader sat on the right side of the vehicle.

Spare road wheels were often fixed to the rear engine deck. The vehicle was fitted with six enlarged 520-mm-diameter road wheels on both sides to cope with the extra weight, not the normal Panzer IV tank's eight pairs of 470-mm-diameter road wheels per side. In the case of damage to the wheels, they could be changed by the crew. A square jacking block was affixed to the right side of the hull just under the rear of the turret. The jack was kept on the rear track guard at the back near to it. Metal tow cables were stored around the outer turret armour plates.

Spare track links were fixed to the rear storage box at the back of the turret. A mock driver's armoured vison slit replaced the hull machine gun on the right side of the front upper hull to confuse enemy gunners. The driver sat on the left of the SPG.

During 1942, the ten prototypes underwent trials on the German Army's eastern test range with the *Feld-Versuchs Batterie* (field test battery), 16th Panzer Artillery Regiment, 12th Panzer Division. These were successful. An order for a further 200 10.5-cm leFH 18/1 (Sf) *auf* GW IVb SPGs was placed. They were going to be built at Stahlindustrie in Mülheim an der Ruhr. The problem was building enough Panzer IV tank chassis to be converted into self-propelled artillery guns. More Panzer IV tanks were urgently required to deal with the Soviet Army's new T-34 and KV-1 tanks. They were the priority.

In the meantime, the engineers realised that they could mount the same 10.5-cm artillery gun on the now obsolete Panzer II tank chassis. This new SPG was later known as the Wespe. No more 10.5-cm leFH 18/1 (Sf) *auf* GW IVb self-propelled artillery guns were built. The order for the additional 200 was cancelled in November 1942.

Active Service

All ten vehicles served with the *Panzer-Artillerie-Regiment* 16, attached to the 16. *Panzerdivision* on the Eastern Front. The division was held in reserve during the Balkan

Campaign, but took part in the invasion of the Soviet Union, Operation Barbarossa, in June 1941.

It was used in the southern sector of the front, advancing to Stalingrad via Lvov, Pervomaisk, Zaporozhe, Taganrog, Makiivka, and Artemovsk. The 10.5-cm leFH 18/1 (Sf) *auf Geschützwagen* IVb joined the rest of 16. *Panzerdivision* in late December early January. The regiment was destroyed in early 1943 in the Stalingrad pocket.

Surviving SPGs

There are no surviving examples of the 10.5-cm leFH 18/1 (Sf) *auf Geschützwagen* IVb (Sd.Kfz.165/1)

Specifications

Length: 5.9 m (19 feet 4 inches)
Width: 2.87 m (9 feet 5 inches)
Height: 2.25 m (7 feet 5 inches)
Total weight, battle ready: 18 tonnes (17.71 tons)
Crew: Four (commander, driver, gunner, and loader)
Prototype engine: Maybach HL 66 P six-cylinder inline water-cooled 6.6-litre gasoline/petrol engine, 188 hp
Production engine: Maybach HL 90 P20k twelve-cylinder water-cooled gasoline/petrol engine, 320 hp
Top road speed: 35 km/h (22 mph)
Operational range on road: 240 km (149 miles)
Operational range off-road: 130 km (81 miles)
Main armament: 10.5-cm leFH 18/1 L/28 howitzer, sixty rounds
Front hull armour: 20 mm
Side and rear hull armour: 14.5 mm
Front turret armour: 20 mm
Side and rear turret armour: 14.5 mm
Total production: Ten (with two prototypes)

10.5-cm leFH 18/1 (Sf) *auf Geschützwagen* IVb (Sd.Kfz.165/1) artillery self-propelled gun. (*David Bocquelet*)

10.5-cm leFH 18/1 (Sf) *auf Geschützwagen* IVb (Sd.Kfz.165/1) artillery self-propelled gun in winter whitewash livery for fighting on the Eastern Front. (*David Bocquelet*)

6

10.5-cm leFH 16 (Sf) *auf Geschützwagen* FCM(f)

The French FCM 36 tank (*Char léger Modèle* 1936 FCM) was a light infantry tank, designed for the French Army prior to the Second World War. They were manufactured by the French company Forges et Chantiers de la Méditerranée (FCM). It had a crew of two and was equipped with a short 3.7-cm main gun and a 7.5-mm coaxial machine gun.

After the Fall of France in 1940, some captured FCM 36 tanks were modified by the Germans to become self-propelled artillery guns. The total amount of 10.5-cm leFH 16 (Sf) *auf Geschützwagen* FCM 36(f) built has not been confirmed. Some say only eight, while other sources say twelve or even forty-eight. At present, there is no documentary evidence to confirm the exact number. The reason why eight is the preferred number is because of a photograph taken inside the tank conversion factory workshop that shows six 10.5cm gun barrels on the floor waiting to be hoisted on to the new built SPG gun mounts and in the background, there are two FCM 36 tank-based artillery SPGs already fitted with their gun barrels.

There is good evidence that twelve were produced as German Army orders show eight being sent to an Artillery Battalion 31 October 1942 and later four more being sent to the same unit in early 1943.

There were a few disadvantages of an open-topped vehicle. The crew was exposed to the elements and were also at risk of injury from enemy-thrown hand grenades, mortars, and shrapnel from air burst enemy shells. Rain covers were produced. They covered both the crew compartment and the gun. The canvas was attached to the upper protective armour using the small D-shaped rings welded to the upper part of the structure.

As the French FCM 36 tank chassis was small, there was limited space for the storage of ammunition. Only thirty-six 10.5-cm HE two-part shells could be carried. The propellant charges were kept on the left of the vehicle while the projectile shells were stored on the right.

A 7.92-mm MG 34 machine gun was attached to top right side of the armour casement, on a swivel mount to the right of the main gun. Drums carrying fifty rounds of spare ammunition were stored underneath the mount. It fired 7.92-mm (0.31-inch) bullets.

Only 100 French-built FCM 36 tanks were completed by the time of the German invasion of France in May 1940. In early 1939, the French Army 4e and 7e Tank Battalions were equipped with fifty tanks apiece. After the German invasion of Poland in September

1939, the two battalions were consolidated under the banner of the 503e French Second Army reserve. The FCM 36 tanks saw active service when they attempted to counter the growing German presence at a bridgehead being set up along the Meuse at Sedan.

With the Fall of France, it is believed that roughly fifty FCM 36 tanks remained in operational service. The Germans decided to use some of these French tanks to help strengthen their occupation forces around France. These captured tanks were known as *Beutepanzers*: trophy tanks. Thirty-seven were used as tanks and give the German Army designation of *Panzerkampfwagen* FCM 737(f). The letter 'f' indicates that the tank was of French origins. Ten FCM 36 tank chassis were used to mount 7.5-cm PaK anti-tank guns. These tank destroyers were known as the Marder I.

It is not clear if *Panzerkampfwagen* FCM 737(f) tanks were withdrawn from internal security patrols and converted into self-propelled artillery and anti-tank guns or if the chassis came from knocked out or abandoned FCM 36 tanks that were recovered or captured on the battlefield.

It was found that the two-part ammunition used on the larger more modern guns, the 10.5-cm leFH 18, could be used up to canister charge five on the First World War-era 10.5-cm leFH 16 light field howitzer. This helped in dealing with the logistics of ammunition supplies.

Operational Deployment

Eight 10.5-cm leFH 16 (Sf) *auf* GW FCM 36(f) self-propelled artillery guns were issued to the *gepanzer Artillerie-Abteilung* (Sf) z.b.V. (Armoured Artillery Battalion) on the Western Front through an order dated 31 October 1942 (K.St.N.430). The letters z.b.V. were the abbreviation for '*zur besonderen Verwendung*'. An English translation would be 'for special deployment' or 'for special assignment'.

They were divided up between two self-propelled artillery batteries called 1. *Batterie* (Sfl) and 2. *Batterie* (Sfl). Four SPGs were in each battery. The letters 'Sfl' stand for *Selbstfahrlafette*, which translates to 'gun carriage' or 'self-propelled gun'. The abbreviation (Sf) is also used.

A further four were issued for deployment. They were put in the 3. *Batterie* (Sf). This gives strength to the argument that twelve 10.5-cm leFH 16 (Sf) *auf* GW FCM 36(f) were built.

On the side of the upper armour of one vehicle in one of the surviving photographs is the number '2' over a '6Gp'. This means that this vehicle was the gun tank number six in the 2nd Battery. The 1st Battery would have SPGs numbered one, two, three, and four. The 2nd Battery would have SPGs numbered five, six, seven, and eight. The 3rd Battery would have SPGs numbered nine, ten, eleven, and twelve.

In March 1943, the unit was renamed, becoming the *Sturmgeschütz-Abteilung* 931 (931st Assault Gun Battalion). This unit's long name is often abbreviated to Stu.Gesch. Abt. 931. This battalion was also equipped with 7.5-cm PaK 40 *auf* FMC 36(f) anti-tank self-propelled guns and an anti-aircraft battery of 2-cm Flak *auf* gep.Zgkw. P 107 vehicles for self-defence.

On 6 May 1943, Stu.Gesch.Abt. 931 was merged with Pz.Jg.Kp. 931 and renamed verst. Pz.Jg.Abt. and was now part of the *Schnelle Division West* (Western fast response division). The abbreviation *'verst'* was for the word *verstärkte*, which means 'reinforced' (reinforced tank hunting battalion).

Pz.Jg.Kp. 931 was equipped with seven gep.Zgkw Somua MCG/MC 7.5-cm PaK 40 *auf* m SPW S307(f) anti-tank self-propelled guns. The abbreviation gep.Zgkw is short for *Gepanzerte Zugkraftwagen* (armoured half-track).

A few weeks later, on 27 June 1943, the unit was renamed again. It was just called *Sturmgeschütz-Abteilung* (Stu.Gesch.Abt.) but without a unit number. The *Schnelle Division West* was now called the 21. *Panzerdivision.*

The rest of the men and equipment of the Pz.Jg.Kp. stayed part of the 21st Panzer Division, but they handed over their gep.Zgkw. Somua *mit* 7.5-cm PaK 40 half-tracks and 2-cm Flak *auf* gep.Zgkw. P 107 to one of the division's *Panzergrenadier* regiments.

On 15 July 1943, it was renamed again. This time, it was called the *Sturmgeschütz-Abteilung* 200 (200th Assault Gun Battalion) and was still part of 21. *Panzerdivision.*

In September 1943, the 4th Battery was equipped with six 105-cm leFH 18 (Sf) *auf Geschützwagen* 39H(f) self-propelled artillery guns. These had artillery guns that could fire high-explosive shells over longer distances. The 4th Battery was also equipped with four 7.5-cm PaK 40 (Sf) *auf* GW 39H(f) anti-tank SPGs. To make the supply of spare parts and maintenance simple, a decision was made in December 1943 that all the all FCM 36 tank chassis-based SPGs were to be replaced with Hotchkiss tank chassis-based SPGs.

German military units regularly sent reports to headquarters on how many soldiers and guns, tanks and SPGs were fit for action. In a battle strength report submitted by this unit dated 1 January 1944, there were no 10.5-cm leFH 16 (Sf) *auf* GW FCM 36(f) listed.

It looks like they did not see action in Normandy fighting off the Allied invasion of France. At present, it is not known what happened to them.

Identification

One of the easiest ways of telling the difference between a 10.5-cm leFH 16 (Sf) *auf Geschützwagen* FCM 36H(f) self-propelled gun and a 7.5-cm PaK 40 (Sf) *auf* PzKpfw FCM 36(f) anti-tank gun SPG is to look at the armoured housing that surrounds the gun's recoil recuperator mechanisms. A recuperator on an artillery gun is a device employing springs or pneumatic power to return a gun to the firing position after the recoil. On the 10.5-cm leFH 16, it is long and is half the length of the gun. It is situated below the gun. The armoured housing covering the 7.5-cm PaK 40 gun's recoil recuperator mechanisms is small and the gun barrel is thinner and much longer.

When looking at different German self-propelled guns, the triangular road wheel armour covering between the tracks is very unique to the FCM 36 tank-based SPGs and makes it easily identifiable.

Surviving SPGs

There are no surviving examples of the 10.5-cm leFH 16 (Sf) *auf Geschützwagen* FCM(f) self-propelled artillery gun.

Specifications

Length: 4.60 m, without gun 4.46 m (15 feet 1 inch, w/g 14 feet 7 inches)
Width: 2.14 m (7 feet)
Height: 2.15 m (7 feet)
Total weight, battle ready: 12.2 tonnes (12 tons)
Crew: Four (commander, driver, gunner, and loader)
Propulsion: Berliet MDP V-4 diesel engine, 91 hp
Fuel capacity: 260 litres
Top road speed: 24 km/h (15 mph)
Operational range (road): 225 km (140 miles)
Main armament: 10.5-cm (4.13 inches) leFH 16 howitzer with thirty-seven rounds
Secondary armament: 7.92-mm (0.31 inches) MG 34 machine gun
Hull armour: Front 25–40 mm (0.98–1.57 inches)
Hull sides and rear armour: 20 mm (0.79 inches)
Upper armour 15 mm (0.59 inches)
Total production: Twelve

10.5-cm leFH 16 *auf Geschützwagen* FCM 36(f) artillery self-propelled gun, having just come out of the tank conversion workshop. (*David Bocquelet*)

10.5-cm leFH 16 *auf Geschützwagen* FCM 36(f) artillery self-propelled gun, 21st Panzer Division, Normandy, summer 1943. (*David Bocquelet*)

10.5-cm leFH 16 *auf Geschützwagen* FCM 36(f) artillery self-propelled gun in winter whitewash livery. (*David Bocquelet*)

7

10.5-cm leFH 18/3 (Sf) *auf Geschützwagen* B2(f)

With the Fall of France in May 1940, the German Army captured a lot of French Army tanks and vehicles. They called them *Beutepanzer* (trophy tanks).

The approximate numbers of French AFVs captured by the German Wehrmacht are as follows: 300 Panhard-178; 3,000 Renault-UE; 350–360 Lorraine 37L and 38L; 1,700 Renault FT Tanks (various versions); 843 Renault R35; 550–600 Hotchkiss H-35 and H-39; 50 FCM-36 tanks; 297 Somua S-35 tanks; and 161 *Char* B1 bis tanks.

More than 5,000 captured French Army armoured fighting vehicles were repaired or converted in French factories. Vehicle manuals were translated into German and the vehicle spare part production initiated.

One such *Beutepanzer* trophy tank was the very large *Char* B1 bis heavy tank. Many of them had been abandoned by their crews after running out of ammunition or fuel. The French Army supply system was poorly organised in 1940. Very few of these monster tanks were knocked out by enemy action.

They were given the German designation *Panzerkampfwagen* B2 740(f) shortened to Pz.Kpfw. B2 740(f). The letter 'f' signified that it was a French-built tank. Hitler issued a directive in March 1941 that sixteen of these tanks were to be converted to self-propelled artillery guns and used to support the *Flammenwerfer auf* Pz.Kpfw.B2(f) flamethrower tanks as they assaulted Soviet positions on the Eastern Front.

It took longer than first planned to convert the Char B1 tanks to self-propelled artillery guns. Lack of equipment, vehicles, guns, and parts were blamed. The *Flammenwerfer auf* Pz.Kpfw. B2(f) flamethrower tanks went into action without supporting artillery SPGs in June 1941.

The *Panzer-Abteilung* (*Flamm*).103 was disbanded by mid-July 1941. The B2 *Flammenwerfer* tanks were a disappointment to the Germans. They kept on breaking down due to mechanical problems. They were not reliable and could not be depended upon to get to an enemy strong point and neutralise it.

Livery

Most of the 10.5-cm leFH 18/3 (Sf) *auf Geschützwagen* B2(f) self-propelled artillery guns were painted '*Panzergrau*' (tank grey) at the factory but a few were repainted in the standard late war paint scheme of sandy yellow d*unkelgelb*.

There is photographic evidence that shows two 10.5-cm leFH 18/3 (Sf) *auf Geschützwagen* B2(f) self-propelled artillery guns painted in two different colours. The first vehicle is painted in the sandy yellow *dunkelgelb* and the one following is painted in the earlier Panzer grey.

Design and Production

The design contract was given to Rheinmetall-Borsig, which was based in Düsseldorf. They created blueprints for a mount placed on top of the tank chassis that could take the 10.5-cm leFH 18/3 artillery howitzer. The gun crew would be protected from small arms fire by an open-topped armoured casement. Its armour plate thickness ranged between 30–40 mm (1.18-1.57 inches) thick at the front and 20–30 mm (0.79-1.18 inches) at the sides. Tarpaulins would cover the open top to give the crew shelter from bad and hot weather.

It had a five-man crew: commander, driver, and three gunners. The *Char* B1 main gun in the hull had been removed. It was not replaced with a hull-mounted machine-gun. For self-defence, a 7.92-mm (0.31-inch) MG 34 machine gun was available to be fitted to the top of the gun casement. The crew also had access to handheld 9-mm (0.35-inch) machine pistols.

The *Char* B1 bis tanks already had a very high profile. With the mounting of the 10.5-cm leFH 18/3 howitzer on the top, it was very tall. Its height was 3 m (9 feet 11 inches). This made it easy for the enemy to spot, so it had to be deployed behind the initial front line of tanks.

The French Army radios were replaced. Some could only send messages by Morse code. Modern German FuG 5 radios that could provide voice communication were fitted in these machines. The removal of the hull-mounted 7.5-cm (2.95-inch) gun increased the amount of room there was available for the storing of extra equipment and ammunition.

Operational Service

Five were produced by January 1942, five in the following February, and six in March. The sixteen 10.5-cm leFH 18/3 (Sf) *auf* G.W. B2(f) were issued to *Artillerie-Regiment* 93, 1. *Abteilung* (1st Battalion) of the 26. *Panzerdivision*, which was in France. The 1st Battalion had three batteries with four self-propelled artillery guns in each battery. The four remaining vehicles were issued to the HQ battery on reserve. Reports stated that the vehicle was prone to frequent mechanical breakdowns.

An *Artillerie-Regiment* 93's combat strength report dated 31 May 1943 reported that they had fifteen 10.5-cm leFH 18/3 (Sf) *auf* G.W. B2(f) SPGs of which fourteen were in an operational condition. This report also recorded that twelve Wespe 10.5-cm SPGs had been issued to the 1.Abt./Artl.Rgt.93 as replacement vehicles for the mechanically unreliable 10.5-cm leFH 18/3 (Sf) *auf Geschützwagen* B2(f) SPGs. These were then used as training vehicles to teach new tank drivers, gunners and mechanics their trade skills. The 26. *Panzerdivision* was sent to Italy in July 1943.

Orders had been issued to send these vehicles to Sardinia to be part of the strength of the 90. *Panzergrenadierdivision*. No records have been found to confirm that the fifteen remaining vehicles were transferred to Sardinia or what their ultimate fate was. The 90. *Panzergrenadierdivision* withdrew to Corsica from Sardinia and was then sent to northern Italy. It fought at Anzio, Rome, the Caesar and Gothic Lines, and the river Po. It was destroyed near Bologna, Italy, in April 1945.

Surviving SPGs

There are no surviving examples of the 10.5-cm leFH 18/3 (Sf) *auf Geschützwagen* B2(f) self-propelled artillery gun.

Specifications

Length: 7.62 m (25 feet)
Width: 2.4 m (7 feet 10 inches)
Height: 3 m (9 feet 11 inches)
Total weight, battle ready: 32.5 tonnes (31.5 tons)
Crew: Five (commander, driver, gunner, and two loaders)
Propulsion: Renault 307 water-cooled, six-cylinder gasoline/petrol engine, 272 hp
Top road speed: 28 km/h (17 mph)
Off road speed: 21 km/h (13 mph)
Operational range (road): 135 km (84 miles)
Main Armament: 10.5-cm (4.13 inches) leFH 18/3 howitzer
Secondary Armament: 7.92-mm (0.31 inches) MG 34 machine gun
Armor: 40–60 mm (1.57–2.36 inches)
Total production: Sixteen

10.5-cm leFH 18/3 (Sf) *auf Geschützwagen* B2(f) artillery self-propelled gun. (*David Bocquelet*)

10.5-cm leFH 18/3 (Sf) *auf Geschützwagen* B2(f) artillery self-propelled gun in dark yellow livery I—a factory fresh condition, with bad weather tarpaulin and gun barrel cover. (*David Bocquelet*)

10.5-cm leFH 18/3 (Sf) *auf Geschützwagen* B2(f) artillery self-propelled gun in *dunkelgelb* livery. (*David Bocquelet*)

8

15-cm sFH 13 (Sf) *auf Geschützwagen* Lorraine Schlepper(f)

A vehicle fitted with an artillery howitzer was called a *Geschützwagen*, which is literally translated as 'gun vehicle'. The letters 'Sf' stand for *Selbstfahrlafette*, meaning self-propelled carriage. The letter (f) indicates that the chassis was of French origin.

Improvised self-propelled artillery guns were developed to enable fast-moving attacks to have artillery support that could keep up with the speed of advancing panzer divisions. They could use direct fire mode at targets they could see or, more commonly, use indirect fire at targets plotted on a map.

They were not designed to be in the front line or engage in combat with tanks. They were motorised artillery guns that could fire high-explosive shells over the heads of friendly troops. Most targets would have been given to the crew as map grid references by forward observation officers or infantry units under attack.

Quite often, the gun crews could not see where their shells landed, as the target was so far away. They would have to rely on the forward observer to tell them if adjustments had to be made.

The open-topped back design of these self-propelled guns had a number of advantages. The elevated commander's position when standing in the crew compartment, behind the protective armoured shield, meant that he had a good view on all sides. If there was the threat of enemy small arms fire, then the crew could use a twin lens range finder telescope that could peak over the top of the armoured casement.

There was enough room for the crew to be transported towards the battlefield while protected from small arms fire and shell shrapnel. The vehicle had good mobility and could follow the infantry almost anywhere. The gun was quicker to get ready for action and fire on targets than towed artillery guns.

They were cheaper and faster to build than a new vehicle. They used the chassis of an obsolete captured French tank and an existing artillery howitzer.

Putting the 15-cm sFH 13/1 howitzer on top of captured French Army Lorraine 37L-tracked armoured supply vehicle chassis was a more efficient use of manpower from the traditional form of German artillery battery transportation. Even in the Second World War, horsepower was still widely used although tracked vehicles were also used when available.

Each field gun would require a six-horse team to pull the gun and limber. The ammunition, supplies and kit would be kept in the limber, which was a very large box on a pair of wheels with seats on the top. Three men would ride on the left-hand horse of each pair to control them. The remaining six men of the gun crew would ride on top of the limber. Only a relative few were towed by the 3-ton half-tracks.

Production

On 23 May 1942, Hitler attended a demonstration of newly constructed self-propelled guns, where captured enemy vehicles had been converted to carry artillery howitzers and anti-tank guns. A decision was made to build 160 *Selbstfahrlafette* (self-propelled guns) based on the Lorraine 37L *Schlepper* tractor. Sixty would carry the 10.5-cm leFH howitzer, forty would carry the 15-cm sFH 13 howitzer, and sixty would have the 7.5-cm PaK 40 anti-tank gun.

Rommel's *Afrika Korps* desperately needed self-propelled artillery guns that could keep up with the tanks. Hitler ordered the immediate assembly of thirty 15-cm sFH 13/1 (Sf) *auf* GW Lorraine Schlepper(f) SPGs on the 25 May 1942. The contract was awarded to Alkett, which were based in Berlin-Borsigwalde. The order was completed in June 1942.

The interior of the vehicle could only hold eight high-explosive shells and eight propellant cartridges. The unusual distinct rear overhanging design was necessary because of the location of the engine. This required a retractable hinged spade to be fitted at the rear of the vehicle. It was lowered into the ground when the gun was in action to help support the vehicle when the gun was fired and stop it moving backwards. The howitzer had a very limited traverse of only five degrees left and right with a 40-degree elevation.

An additional seventy-two were built in France between July and August 1942 by *Baukommando* Becker, bringing the total made in the Second World War to 102. Some sources, that at present cannot be verified, state that a further sixty-four were built by Alkett in their factory near Berlin. That would bring the total built to 166. If one studies the operational photos, it will be noticed that there are three distinct models that have different build features. This would match with those sources which state that Alkett built the first batch of thirty, then an improved version of sixty-four, while *Baukommando* Becker built seventy-two slightly different ones in France.

What are the differences? The first thirty that were sent to the desert had short retractable tail spades at the rear and no metal D-shaped handles mounted on the side and bottom of the front armour plate. The second batch had the short retractable tail spades at the rear and metal D-shaped handles fitted to the side and bottom of the front armour plate. This version can be seen in the photographs of the sixty vehicles on parade in the Palace of Versailles. The third version had extended long retractable tail spades at the rear of the vehicle.

Operational Service

The initial batch of thirty Alkett built 15-cm sFH 13/1 (Sf) *auf* GW Lorraine Schlepper(f) sent to North Africa was divided up between three different panzer divisions. Twelve were going to be sent to the 21st Panzer Division, another twelve were going to be sent to the 15th Panzer Division, and the remaining six were to be issued to the 90th Leicht Division.

Unfortunately for the seven, three were sunk in transit in July 1942 and four more on 4 August. The remaining twenty-three vehicles arrived safely at the Libyan ports of Benghazi and Tobruk. They first saw action during Rommel's final attempt to break through the El Alamein defences on 30 August 1942. The 15th Panzer Division reported that three of their 15-cm sFH 13/1 (Sf) *auf* GW Lorraine Schlepper(f) were total losses during the period from 30 August to 3 September 1942. The remaining nineteen 15-cm self-propelled guns were reported as available on 23 October 1942, the day the British launched their attack on the German defences at El Alamein. All were reported as having been lost by 2 December 1942. Some were captured by the British.

The new batch of sixty 15-cm sFH 13/1 (Sf) *auf* GW Lorraine Schlepper(f) was built in France in the summer of 1942 by *Baukommando* Becker for the *Schnelle* (quick) *Brigade West*. They differed only minimally from the original Alkett built vehicles, with details like the trail spade and travel lock. They were quickly issued to front-line units. Thirty were sent to the *Gepanzerte Artillerie-Regiment* 1 (Sf) and thirty were delivered to the *Gepanzerte Artillerie-Regiment* 2 (Sf). Each regiment comprised of five batteries with six self-propelled artillery guns in each battery.

The *Gepanzerte Artillerie-Regiment* 1 (Sf) was disbanded in December 1942, their vehicles scattered at three apiece to infantry divisions stationed in the west. The *Gepanzerte Artillerie-Regiment* 2 (Sf) was reorganised and renamed *Artillerie-Regiment* 931 in March 1943 and later *Panzer-Artillerie-Regiment* 155, which was part of the 21. *Panzerdivision* (*neu*). By June 1944, regimental documentation showed it only had twelve of the original thirty 15-cm sFH 13/1 SPGs it was issued in 1942.

Other tracked weapons had replaced them and the Regiment now had ten batteries including a rocket battalion. They went into action in Normandy in June 1944. Six were issued to the 6th Battery, 2nd Battalion (II. *Abteilung* (Sf)) and six were issued to the 9th Battery, 3rd Battalion (III. *Abteilung* (Sf)). The *Panzergrenadier Regiment* 125 was shown as having six 15-cm sFH 13/1 SPGs and the *Panzer-Grenadier-Regiment* 192 also had six.

The last ones were knocked out when they were caught in the Falaise Pocket and subjected to intense bombing, shelling and gunfire in August 1944. There are only two original surviving examples left. A third was recently destroyed in Iraq. The best example of a surviving 15-cm sFH 13/1 (Sf) *auf Geschützwagen* Lorraine Schlepper(f) SdKfz 135/1 self-propelled artillery gun was captured at El Alamein and is now kept at the US Army Artillery Museum, Fort Sill, OK, USA. The second is damaged but is on display at the El Alamein War Museum in Egypt.

Identification

One of the easiest ways of telling the difference between a 10.5-cm leFH 18/40 *auf Geschützwagen* Lorraine Schlepper(f) self-propelled artillery gun and the 15-cm sFH 13/1 (Sf) *auf Geschützwagen* Lorraine Schlepper(f) is to look at the armoured housing that surrounds the gun's recoil management recuperator mechanisms.

On the 10.5-cm leFH 18, there are two: one above and below the gun. On the 15-cm sFH (Sf), there is only one below the gun barrel. The 15-cm sFH (Sf) gun was not fitted with a muzzle brake. It does have two metal bands fitted around the barrel, unlike the 10.5-cm leFH-18/40 gun. Both guns were fitted to the same Lorraine 37L tractor chassis.

Panzer-Artillerie-Regiment 155, 21st Panzer Division Reported Strength in June 1944

I. *Abteilung* (1st Battalion)
  1st Battery 4 × 12.2-cm *Kanone* 390/1(r) (captured Russian)
  2nd Battery 4 × 12.2-cm *Kanone* 390/1(r) (captured Russian)
  3rd Battery 4 × 10-cm K 18s
II. *Abteilung* (Sf) (2nd Battalion)
  4th Battery 6 × 10.5-cm leFH 18 *auf* Lorraine
  5th Battery 6 × 10.5-cm leFH 18 *auf* Lorraine
  6th Battery 6 × 15-cm sFH 13 *auf* Lorraine
III. *Abteilung* (Sf) (3rd Battalion)
  7th Battery 6 × 10.5-cm leFH 18 *auf* Lorraine
  8th Battery 6 × 10.5-cm leFH 18 *auf* Lorraine
  9th Battery 6 × 15-cm sFH 13 *auf* Lorraine
Rocket Battalion (*Werfer*)
  10. *Batterie* (10th Battery) 2 × S307(f) *R-Vielfachwerfer*
Kampfgruppe Rauch Panzergrenadier-Regiment 192
  9.(SiG) *Kompanie*: 3 × 15-cm sFH 13 *auf* Lorraine
Panzer-Artillerie-Regiment 155 2.Battalion
  4th Battery 3 × 10.5-cm leFH 18 *auf* Lorraine
  5th Battery 3 × 10.5-cm leFH 18 *auf* Lorraine
  6th Battery 3 × 15-cm sFH 13 *auf* Lorraine
Kampfgruppe Oppeln Panzer-Regiment 22 3.Battalion
  7th Battery 3 × 10.5-cm leFH 18 *auf* Lorraine
  8th Battery 3 × 10.5-cm leFH 18 *auf* Lorraine
  9th Battery 3 × 15-cm sFH 13 *auf* Lorraine
Kampfgruppe Luck Panzergrenadier-Regiment 125
  9.(SiG) *Kompanie*: 3 × 15-cm sFH 13 *auf* Lorraine
1.*Panzergrenadier-Abteilung* (SPW)
  HQ 1 × 15-cm sFH 13 *auf* Lorraine
  9.(SiG) *Kompanie* 3 × 15-cm sFH 13 *auf* Lorraine

Surviving SPGs

US Army Artillery Museum, Fort Sill, OK, USA
El Alamein War Museum, Egypt
Wreck, unknown location, Iraq

Specifications

Length: 4.40 m (without gun) (14 feet 6 inches) 5.31 m with gun (17 feet 5 inches)
Width: 1.85 m (5 feet 1 inch)
Height: 2.02 m (6 feet 7 inches)
Total weight, battle ready: 8.5 tonnes (8.36 tons)
Crew: Four (commander, driver, gunner, and loader)
Propulsion: Type 135 Delahaye six-cylinder water-cooled inline 3.56-litre petrol engine, 70 hp at 2,800 rpm
Fuel capacity: 110 litres
Top speed: 35 km/h (22 mph)
Operational range (road): 120 km (74.5 miles)
Main armament: 15-cm leFH 18 howitzer with thirty-six rounds
Armor: 9–14.5 mm
Total production: 166

Above: Alkett-built 15-cm sFH 13/1 (Sf) *auf* GW Lorraine Schlepper(f) (SdKfz 135/1) artillery self-propelled gun in plain grey livery. Late 1942. (*David Bocquelet*)

Opposite above: Alkett-built 15-cm sFH 13/1 (Sf) *auf* GW Lorraine Schlepper(f) (SdKfz 135/1) artillery self-propelled gun in *dunkelgelb* (dark yellow) and green livery, North Africa, 1942. (*David Bocquelet*)

Opposite middle: Alkett-built 15-cm sFH 13/1 (Sf) *auf* GW Lorraine Schlepper(f) (SdKfz 135/1) artillery self-propelled gun in *dunkelgelb*, brown and green livery. Normandy, summer 1944. (*David Bocquelet*)

Opposite below: Baukommando Becker-built 15-cm sFH 13/1 (Sf) *auf* GW Lorraine Schlepper(f) (SdKfz 135/1) artillery self-propelled gun in *dunkelgelb*, green, and brown livery with long tail spade. Normandy, summer 1944. (*David Bocquelet*)

9

15-cm sIG 33 (Sf) *auf Fahrgestell Panzerkampfwagen* II

In 1941, Field Marshal Erwin Rommel needed heavy artillery to support his advancing armoured divisions in North Africa. Horse-drawn artillery was impractical in the heat of the desert. The logistics of supplying these animals with enough feed and water was a nightmare. He demanded a solution be found.

The Panzer II tank had performed well in Poland in 1939, but it was outclassed by some of the British and French tanks it encountered during the invasion of France in May 1940. This vehicle was considered obsolete by 1941 and was replaced by Panzer III tanks as soon as they came off the factory production line.

The German Army now had a surplus of Panzer II tanks. The decision was taken to remove the tank's turret and mount in its place a 15-cm sIG 33 heavy infantry howitzer with a gun shield to protect the crew. Rather than having to find a crew of nine men and six horses to operate one artillery gun, the German Army only needed to find four crewmen. It is sometimes called one of 'Rommel's Funnies'.

The Name: Not *Sturmpanzer* II or Bison II

The correct German Army designation for this self-propelled artillery gun is 15-cm sIG 33 *auf Fahrgestell Panzerkampfwagen* II (Sf) Sd.Kfz.121/122 or 15-cm s.l.G.33 *auf Fahrgestell* Pz.Kpfw.ll (Sf).

While undergoing trials, it was sometimes referred to as the 15-cm sIG 33 B (this was to show it was an upgrade from the prototype SPG, the 15-cm sIG 33 A, that used the original five wheeled Panzer II tank chassis and had not yet been lengthened).

After the Second World War, a scale model kit company produced one of the first retail kits of this vehicle. They called it the 'Bison II', believing it to be the natural progression for the earlier 15-cm sIG 33 L/12 (Sf) *auf Geschützwagen* I Ausf. B 'Bison' self-propelled artillery gun based on a Panzer I tank chassis.

This was wrong. It was never called the Bison II during the war but, after the war, the name Bison II stuck. Museums, historical books, and other scale model kit companies continue to call this mobile artillery weapon the Bison II.

Some authors, museums, and scale model kit companies also wrongly call it the *Sturmpanzer* II. A '*Sturmpanzer*' is a heavily armoured assault tank. Its crew is protected in a fully armoured box that is enclosed on all four sides and has an armoured roof. The front armour of this vehicle is thick to enable it to get close to enemy strong points before firing its weapon.

This description bears no resemblance to the function and appearance of the German 15-cm sIG 33 *auf Fahrgestell Panzerkampfwagen* II (Sf) SPG. This vehicle was not designed to advance towards heavily fortified gun emplacements and blow them up while under heavy fire. It is only thinly armoured and the crew have very little protection.

This vehicle was designed to keep up with the advancing infantry and tanks, but remain behind them, out of harm's way, and fire shells over their heads at enemy targets.

Production and Development

Before the first 15-cm sIG 33 *auf Fahrgestell Panzerkampfwagen* I (Sf) (Bison I) had been tested in combat in May 1940, the German Army *Waffenamt* was already engaged in the development of an improved model based on the Panzer II Ausf. B tank chassis, rather than the Panzer I.

The prototype was successfully test fired at Kummersdorf on 13 June 1940. A 15-cm sIG 33 B, without standard gun carriage wheels, had been fitted to a lengthened Panzer II tank chassis by Alkett, an engineering company near Berlin. As part of the trials, 120 rounds were fired.

The longer Panzer II tank chassis absorbed the 15-cm gun's highest recoil force of 9 tons firing at all settings and at a rapid rate of fire of up to three rounds per minute. The tracks did not noticeably sink into the ground while firing on sandy soil.

Several months earlier, a similar trial had been conducted by Alkett. The standard field gun carriage wheels had been mounted on a standard Panzer tank chassis. This meant it had to be mounted too far to the rear.

Firing at the highest muzzle velocity of 240 meters/second at an elevation of about 15 degrees, the recoil force of 9 tons passed through the middle of the last wheels, resulting in the Pz.Kpfw.II tank chassis unacceptably tipping backwards.

After this initial failure, the Alkett design team went back to the drawing board. They took the wheels off the gun carriage and devised new mounts. This meant that the 15-cm howitzer could be fitted considerably farther forward on the tank chassis, which made the vehicle significantly more stable when firing.

The *Entwicklungsstück* (prototype development piece) was completed by Alkett in October 1940 using a normal length Pz.Kpfw.II chassis with five road-wheels. They only made one. On 18 February 1941, General Halder commented in his diary: 'Room inside the sIG *auf* Pz.II is unacceptably small. New vehicles in four months. Mass production in one year'.

Alkett must have taken this criticism to heart because the *Versuchsserie* (trial series) of twelve 15-cm s.l.G.33 *auf Fahrgestell* Pz.Kpfw.ll (Sf) SPGs were produced with a lengthened and widened hull resulting in the addition of a sixth road wheel. Delivery was to start in August 1941. Overall length of the new vehicle was 5.48 m, width 2.60 m, and height 1.98 m (compared to 4.81 m long by 2.28 m wide of the Panzer II Ausf. F light tank).

Additional space for the fighting compartment was also achieved by mounting the 150 metric hp Buessing-NAG L8V water-cooled gasoline petrol engine transversely across the rear. The gun had a 75-degree elevation, but only ten rounds could be carried onboard.

Alkett completed seven 15-cm sIG 33 *auf Fahrgestell Panzerkampfwagen* II (Sf) in December 1941 and five in January 1942.

North Africa

On 15 September 1941, the German Army Organisations-*Abteilung* ordered the creation of two new independent heavy self-propelled infantry gun companies: 707.*schwere Infanteriegeschütze-Kompanien* and 708.*schwere Infanteriegeschütze-Kompanien*. These units were sometimes abbreviated to sIG Kp. (mot. S).

Each company were issued with six of these new SPGs. The company was subdivided into three platoons. Each platoon had two SPGs. The six vehicles in the 707 were assigned to the *Schützen-Regiment* 155 and the six vehicles in the 708 were assigned to the *Schützen-Regiment* 200. Both of these regiments were part of Rommel's 90.*leicht Afrika-Division* and fought until the Axis surrender in Tunisia in May 1943. The twelve vehicles did get to Libya in time for Rommel's assault on Tobruk, due to its postponement from November 1941 until June 1942.

On 23 February 1942, the six vehicles belonging to 708 arrived in Tripoli. Three different ships were used, each carrying just two of the SPGs, just in case of a torpedo or bombing attack. The sIG Kp.707 was not transported to Libya until April 1942. Six different ships were used to transport the remaining six vehicles belonging to 707. The last vehicle was taken off the ship in Tripoli on 24 April 1941.

On 9 March, the 708 company commander reported at the division command post that they had arrived in Bengasi on 8 March, but they must remain there for two days to complete various engine repairs. Not enough spare parts had been shipped to North Africa. One of the unit's NCOs flew back to Alkett in Berlin to arrange for replacement parts to be shipped out as soon as possible.

Mechanical problems were going to plague the deployment of these new self-propelled artillery guns in the desert. The engine had not been upgraded to take the additional weight and cope with the conditions encountered in Libya, Egypt, and Tunisia. The Panzer II 155-hp engine was too weak for the vehicle weight of 16 tons.

The engine cooling system was inefficient for conditions in Africa. In accordance with the driver's manual, the cooler flaps must be opened while driving. The large amount of dust thrown up by the tracks fell through the open grating and plugged up parts of the radiator reducing the available surface area and its ability to cool the radiator fluid. Engine temperatures became too hot.

A report dated 16 May 1942 stated that, 'During the 70 km journey to Segnali-Sued it soon became apparent that the new 15-cm sIG 33 *auf Fahrgestell Panzerkampfwagen* II (Sf) of 708.*schwere Infanteriegeschütze-Kompanien* were unusable due to design faults. The same situation applied to the vehicles of 707. They were currently in Bengasi with two broken-down SPG's without any resources to transport them to the front or repair them'.

The transmission and transfer boxes suffered from faults. After a very short running time, sometimes fifteen minutes, the engines were so hot that they had to stop to cool them. The pressurised engine coolant reached 120 degrees Celsius after a short running time. Cracks appeared in the track and drive wheels due in part to poor material quality.

Panzerarmee Afrika still listed eight 15-cm sIG 33 *auf Fahrgestell Panzerkampfwagen* II (Sf) as available, but not how many were operational on 23 October. All eight were reported as having been lost by 2 December 1942. The British reported capturing six of them abandoned during the withdrawal, all in a dismantled condition in a tank workshop.

Rain and Sun Protection

The crew could fit two metal bands across the middle of the open-topped fighting compartment. They provided similar support as tent poles in a tent. A large canvas tarpaulin tied down at the back of the vehicle could be hauled over these bands and lashed down to the front of the vehicle to give the crew protection from the sun, as well as rain and sandstorms.

Why Were Only Twelve Built?

A German Army *Waffenamt* liaison officer assigned to *Panzerarmee Afrika* filed a report dated 30 August 1942 about the use of the 15-cm sIG 33 *auf Fahrgestell Panzerkampfwagen* II (Sf) in North Africa:

> From the ballistic and destructive effectiveness standpoint, the gun is judged to be outstanding but the Panzer II chassis is too underpowered and liable to mechanical breakdown. The troops propose that a more suitable tank chassis should be used. I do not agree. The range of this gun cannot be increased. The expenditure is not justified. It would be better to use the 15-cm sFH13 Sfl heavy field howitzer, which has the same calibre but a considerably longer range, on a better tank chassis.
>
> Operationally it is too risky to employ such a valuable self-propelled gun as a fixed location defensive weapon. As soon as it opens fire the gun will be knocked out by concentrated artillery fire. In this situation, they should be replaced with a 12-cm *Werfer* (mortar) that, with its increased rate of fire, lower weight and easier transport on a truck, has almost the same destructive effectiveness.

Lessons learned from this vehicle, the 15-cm sIG 33 *auf Fahrgestell Panzerkampfwagen* II (Sf) Sd.Kfz.121/122, and earlier 15-cm and 10.5-cm self-propelled artillery guns resulted in the design and development of the 10.5-cm Wespe self-propelled artillery infantry gun.

Surviving SPGs

There are no surviving examples of the 15-cm sIG 33 (Sf) *auf Fahrgestell Panzerkampfwagen* II self-propelled artillery guns.

Specifications

Length: 5.48 m (17 feet 11 inches)
Width: 2.60 m (8 feet 6 inches)
Height: 1.98 m (6 feet 6 inches)
Total weight, battle ready: 12 tonnes (11.81 tons)
Armament: 15-cm sIG 33 L/11 howitzer, ten rounds
Front hull armour: 30 mm (1.18 inches)
Superstructure armour: 15 mm (0.59 inches)
Crew: Four (driver, commander, gunner, and loader)
Propulsion: Buessing-NAG L8V water-cooled gasoline/petrol engine
Max Road Speed: 45 km/h (28 mph)
Range: 100 km (110 miles)
Total production: Twelve

Above: The prototype 15-cm sIG 33 A artillery self-propelled gun only had five wheels and the normal length Panzer II tank chassis. (*David Bocquelet*)

Opposite above: The 15-cm sIG 33 *auf Fahrgestell Panzerkampfwagen* II (Sf) artillery self-propelled gun in grey livery. (*David Bocquelet*)

Opposite middle: 15-cm sIG 33 *auf Fahrgestell Panzerkampfwagen* II (Sf) artillery self-propelled gun in *dunkelgelb* livery ready for active service in North Africa, 1941. (*David Bocquelet*)

Opposite below: 15-cm sIG 33 *auf Fahrgestell Panzerkampfwagen* II (Sf) artillery self-propelled gun with tarpaulin frame erected and sunshade tarpaulin tied on. (*David Bocquelet*)

10
15-cm sIG 33 L/11 (Sf) *auf Fahrgestell Panzerkampfwagen* III Ausf. H

Troops are very resourceful and often make battlefield modifications to existing equipment. One such example was the 15-cm sIG 33 L/11(Sf) *auf Fahrgestell Panzerkampfwagen* III Ausf. H (Sf). Only one was ever made and it did not survive the war. Luckily, photographs are available to prove its existence and use in North Africa.

A Panzer III Ausf. H tank chassis was used to mount the same 15-cm sIG 33 L/11 heavy field howitzer that was used on the 15-cm sIG 33 *auf Fahrgestell Panzerkampfwagen* II (Sf) that had an extended Panzer II tank chassis.

This self-propelled artillery gun also has a problem with its name. Some scale model kit companies, military history authors, and museums wrongly call the field modification 15-cm sIG 33 L/11 *auf Fahrgestell Panzerkampfwagen* III Ausf. H (Sf) SPG the Bison III. It was never called that during the Second World War.

It is believed that it was used by the German *Afrika Korps Schützen-Regiment*.200 (90 *leichte Infanterie-Division*) attached to the sIG *Kompanie* (Sf) 708 between 1942 and 1943 and first saw action in September 1942.

Why Was It Built?

In 1941, Field Marshal Erwin Rommel needed heavy artillery to support his advancing armoured divisions in North Africa. Horse-drawn artillery was impractical in the heat of the desert. The logistics of supplying these animals with enough feed and water was a nightmare. Each towed artillery gun would require a team of six horses to pull it.

He demanded a solution be found. A small number of 10.5-cm and 15-cm artillery field howitzers were mounted on captured enemy tanks and tractor chassis and sent to support the men and machines of the German *Afrika Korps* (DAK). Some were fitted to obsolete German tank chassis like the Panzer I and Panzer II. These tanks were replaced by the Panzer III and Panzer IV as they came off the factory production line and entered service.

These self-propelled German artillery guns are sometimes referred to as 'Rommel's Funnies'. This vehicle was not designed to advance towards heavily fortified gun emplacements and blow them up while under heavy fire. It has only a thinly armoured fighting compartment: the crew had very little protection. It was intended to keep up with the advancing infantry and tanks but remained behind them out of harm's way and fire shells over their heads at enemy targets.

Why Was Only One Built?

When vehicles get damaged on the battlefield, they are recovered by the engineers who try to repair them back behind the front line. Sometimes, parts of an armoured fighting vehicle are damaged beyond repair but the rest of the vehicle is in good working order. The *Afrika Korps* engineers had problems getting replacement parts in North Africa because of the sinking of Axis transport ships in the Mediterranean by the Allies. Rommel's forces were outnumbered by the Allies in North Africa who had more tanks, artillery guns, and armoured fighting vehicles.

The German engineers were under pressure to repair and return as many serviceable fighting vehicles to the front line as they could. It would have seemed logical to them that a working 15-cm sIG heavy infantry gun salvaged from a knocked out self-propelled gun should be sent back to the front on a new chassis. In this situation, they used a turret-less recovered panzer III Ausf. H tank chassis. No more were built because development of the 15-cm sFH 18/1 (Sf) *auf Geschützwagen* III/IV Hummel (Sd.Kfz.165) took priority.

Surviving SPGs

There are no surviving examples of the 15-cm sIG 33 L/11 (Sf) *auf Fahrgestell Panzerkampfwagen* III Ausf. H self-propelled artillery guns.

Specifications

Length: 5.41 m (17 feet 9 inches)
Width: 2.95 m (9 feet 8 inches)
Height: 2.44 m (8 feet)
Total weight, battle ready: 20.3 tonnes (19.97 tons)
Crew: Five
Propulsion: Maybach V12 gasoline HL 120 TRM (220 kW) 300 bhp
Speed on /off road: 40/20 km/h (25/12 mph)
Main Armament: 15-cm sIG 33 L/11 heavy infantry gun
Secondary Armament: 7.92-mm *Maschinengewehr* 34 machine gun
Armour: 16–30 mm
Range: 165 km (102 mi)
Total production: 5,774

15-cm sIG 33 L/11 (Sf) *auf Fahrgestell Panzerkampfwagen* III Ausf. H artillery self-propelled gun, one of 'Rommel's Funnies'. (*David Bocquelet*)

Above: 15-cm sIG 33 L/11 (Sf) *auf Fahrgestell Panzerkampfwagen* III Ausf. H artillery self-propelled gun with gun in forward position for travelling. (*David Bocquelet*)

Below: 15-cm sIG 33 L/11 (Sf) *auf Fahrgestell Panzerkampfwagen* III Ausf. H artillery self-propelled gun with bad weather tarpaulin. (*David Bocquelet*)

11

10.5-cm leFH 18/2 (Sf) *auf Fahrgestell Panzerkampfwagen* II Wespe (Sd.Kfz.124)

The Wespe was a German Army artillery 10.5-cm self-propelled gun based on the Panzer II light tank (Sd.Kfz.121) chassis. Its full designation was 10.5-cm *Leichte Feldhaubitze* 18/2 *auf Fahrgestell Panzerkampfwagen* II *Selbstfahrlafette* (Sd.Kfz. 124), which was abbreviated to 10.5-cm LeFH 18/2 *auf* Fgst. PzKpfw II (Sf). It was given the nickname '*Wespe*': German for wasp. This self-propelled gun could fire a 10.5-cm fragmentation high-explosive shells at enemy targets 13.25 km (8.23 miles) away. The term '*Fahrgestell Panzerkampfwagen* II' translates to 'armoured combat vehicle chassis' or 'Panzer II tank chassis'.

The Panzer II light tank was used in the invasion of France in May 1940. It was unable to knock out many of the heavily armoured Allied tanks like the Matilda II or the French *Char* B1 *bis*. Its armour was thin and could be penetrated by the British 2-pounder and French 47-mm SA 35 anti-tank gun. The Panzer II was withdrawn from front-line active service.

Alkett's design team converted the obsolete Panzer II tank chassis into tank destroyers. The turret was removed and a 75-mm PaK 40 anti-tank gun was fitted, surrounded by an armoured open-topped superstructure. They were known as the Marder II (Sd.Kfz. 131). Later, captured Soviet 7.62-mm anti-tank guns were fitted on top of Panzer II light tank chassis. They were called Marder II (Sd.Kfz. 132)

This was a quick way of fitting a large gun on to a tracked vehicle that could keep up with the panzer divisions as they crossed rough muddy open countryside of Poland and Russia during Operation Barbarossa, the invasion of the Soviet Union, in June 1941. At that time, a solution had not been found on how to fit a 75-mm anti-tank gun into a tank turret.

The advancing German Army also needed artillery support. As the rains of autumn and then the snows of winter began to take hold of the battlefield, towing artillery guns became problematic. Horse-drawn artillery guns and towed artillery guns soon got stuck in the mud or snow. Tracked self-propelled artillery guns could cross this testing terrain and keep up with the troops and tanks. The Army wanted self-propelled artillery guns. German industry came up with a number of different solutions, one of which was the Wespe.

Design and Production

Alkett's engineers choose the Panzer II Ausf. F light tank chassis for the conversion. They were conscious of the large recoil that the tank chassis would have to absorb during the firing of this large powerful 10.5-cm artillery gun. The original Panzer II tank was only armed with a 20-mm KwK 30/38 gun.

Alkett's designers relocated the engine from the rear of the tank chassis to the middle of the hull. Air intakes were fitted on either side. This enabled the fighting compartment to be shifted to the rear. This enabled the gun crew more room to work the gun. The transmission system was kept in its original location next to the driver on his right.

Unlike other self-propelled artillery conversions, rear 'trails and spades' were not fitted to the rear of the vehicle. These legs were normally lowered at the rear of the SPG and hammered into the ground to provide a better anchorage when the gun was fired. The Alkett designers felt this was not necessary on the Wespe artillery SPG.

The tank chassis was lengthened at the rear to make room for the Rheinmetall-Borsig 10.5-cm leFH 18/2 L/28 light field howitzer. It was mounted in the centre of the open-topped armoured superstructure over the engine and behind a gun shield. The gun could be traversed 17 degrees left and right. It could be depressed 5 degrees and elevated 42 degrees.

The superstructure built around the gun fighting compartment was only 10 mm (0.39 inches) thick. Like other self-propelled guns of this period, it had a low rear and open-topped configuration. It would only protect the crew from small arms fire and fragmentation high-explosive shell shrapnel. It would not protect them from an armour-piercing round. This vehicle was not meant to be on the front line engaging enemy tanks. It was a support weapon that fired over the heads of its own troops and tanks from behind the front line. The driver's compartment was at the front of the vehicle. It was fully enclosed in armour.

During inclement weather and scorching summer days, a tarpaulin could be secured over the open top to give the crew some protection from rain, snow, and some shade from the scorching sun.

The Wespe artillery SPG was powered by a Maybach HL 62 TRM 140 PS six-cylinder petrol 138-hp (103 kW) engine fed by two 140-l fuel tanks. It had an operational range on the road of around 200 km (120 miles) before it needed refuelling. The ZFA SSG 46 Aphon gearbox provided one reverse gear and six forward gears.

The Panzer II light tank weighted 8.9 tonnes (8.8 tons) but the weight of the Wespe artillery SPG rose to 11 tonnes (10.82 tons). This put additional stress on the engine but it seemed to cope with the weight increase as it still had a reported maximum speed of 40 km/h (24.85 mph). Additional truncated cone springs were added to the first and second road wheels on both sides of the chassis to reinforce the suspension to help the vehicle absorb the gun's recoil and additional weight. The amount of track return rollers each side was reduced from four to three.

The frontal glacis armoured plate on the tank chassis was redesigned. The driver was given an armoured compartment on the left side of the tank that protruded through the slopping front armour. A tubular bullet deflector was welded to the front armour in front of the driver's cabin and vision slot. The driver could only see to his left or right through small slits in the armour of his cabin. The shallow angle of the front armour gave greater protection by increasing the thickness of metal any incoming round would have to penetrate. It also provided a greater chance of a ricochet.

The driver climbed into his position through a two-piece hatch which opened forward and back over his seat. The engine in the middle of the tank chassis separated the driver from the gun crew fighting compartment. The driver could access the engine compartment via a sliding door behind his seat. There was a circular service hatch in the front armour on the right side of the vehicle that could be unscrewed to allow access to the transmission and steering components.

Alkett's prototype Wespe was produced at their Berlin-Borsigwalde factory. On 14 July 1942, after the prototype successfully completed trials, the *Panzer Kommission* recommended that the Wespe was fit for purpose. On 25 July 1942, Hitler confirmed an order for the production of 1,000 10.5-cm leFH 18/2 (Sf) *auf Fahrgestell Panzerkampfwagen* II *Wespe* (Sd.Kfz.124).

Two factories in Poland were involved in building the Wespe artillery SPG: Ursus–Fahrzeug *und* Motorenbau (FAMO) of Breslau (Wroclaw) and Vereinigte *Maschinenwerke*/Famo Warschau in Warsaw. FAMO had previously been involved in building the Panzer II Ausf. F light tank, the vehicle that would provide the chassis for the Wespe. The modification of the 10.5-cm leFH 18 light field gun to enable it to be mounted on top of the extended Panzer II tank chassis was carried out by the gun's manufacturer Rheinmetall-Borsig of Düsseldorf. Records show that 676 were manufactured. The start of production was delayed until February 1943. It stopped in June 1944, when the Soviet Army took control of the area.

To enable communications with other Wespes in the same battery, a FuG Spr.f USW radio transmitter and receiver fitted inside the hull on the left side of the superstructure in a special bespoke frame fitted with rubber shock absorbers. Command vehicles could be identified by an additional aerial fitted at the rear, the top of which resembled an umbrella that had been blown inside out and lost its canvas. They had an additional FuG 7 radio set to communicate with HQ, artillery observes and other units.

For local self-defence against enemy infantry or aircraft, the crew could mount a 7.92-mm *Maschinengewehr* 34, (MG 34) or MG 40 recoil-operated air-cooled machine gun on the front edge of the superstructure. If they were surprised by an enemy tank they could fire an armour-piercing shell but it was only effective at ranges of 500 m or below. They also had access to handheld MP 38 or MP 40 sub machine guns as well as their personal rifles or handguns.

On top of the right track guard holders were fitted to take an axe, 'S' hook, jack resting block, jack, and clamps for a wire cutter. On the left track guard, there were holders for a Bosch headlight, pickaxe, track tensioning tool, towing cable and shovel. The thin metal tube that links the three elements of the gun is part of the gun locking system. When the SPG was driving across country, the gun barrel could be clamped into position to prevent any damage to it. When the gun was needed for firing, the lock would be released. There was also a tall clamp in the fighting compartment that was used to hold the rear of the gun in position.

Operational Service

In March 1943, Eastern Front *Panzer-Artillerie-Abteilungen* (armoured artillery battalions) started to be equipped with the 10.5-cm Wespe SPGs. They first saw action alongside the 15-cm Hummel artillery SPGs during Operation *Zitadelle* (the Battle of Kursk). Later, they were issued to all three Eastern Front army groups.

The Wespe SPGs left the factory painted in a *dunkelgelb* base colour. When they arrived near the front line, the gun crews or field maintenance units painted them with a variety of different camouflage schemes to help the vehicles merge with the local vegetation. *Olivegrun* (green) and *rotbrun* (red-brown) paint was applied using an airbrush but if this was not available, it was painted on the sides of the Wespe SPG using a paintbrush, sponge, cloth, or broom. Some regiments used *panzergrau* (tank grey) paint in their camouflage schemes. During the winter, whitewash was painted over the vehicles so they merged with the snow.

Battlefield reports were favourable. This made Hitler order all other Panzer II tank chassis-based conversions to be stopped. He insisted that all efforts should go towards building as many Wespe self-propelled artillery guns as possible.

Wespe SPGs were used in Italy to provide artillery bombardment of the Allied forces defending the Caesar and Gustav lines. They provided support for the German troops fighting the allies in the Anzio pocket.

By the summer of 1944, some Wespe equipped *Panzer-Artillerie-Abteilungen* were sent to Normandy as a strengthening measure to deal with the expected Allied invasion.

In December 1944, Wespe SPGs were in action during the Battle of the Bulge (Ardennes Offensive). The *Panzer Artillery Regiment* of the Heer Panzer Division had three *Abteilungen* (Battalions). The second and third battalion comprised of towed 10.5-cm, 15-cm, and 17-cm howitzers but the first battalion was equipped with self-propelled artillery guns: twelve Wespe and six Hummel artillery SPGs.

1.*Abteilung* (1st Battalion)
Stabskompanie (HQ company)
1.*Batterie* (6 × Wespe 10.5-cm Artillery SPG)
2.*Batterie* (6 × Wespe 10.5-cm Artillery SPG)
3.*Batterie* (6 × Hummel 15-cm Artillery SPG)

Each battery normally had two *Panzerbeobachtungswagen* armoured artillery observation vehicles. These were often based on tank chassis with the main gun removed to provide additional space for maps and radios. A dummy gun was welded in its place so that the enemy would think it was a normal gun tank. Panzer II to V tanks were used as *Panzerbeobachtungswagens.*

No fewer than thirty-six panzer divisions, including SS and special units, received Wespe SPGs. They saw active service on all battlefronts after 1943. By March 1945, 307 Wespe SPGs (under half the original production number) were reported to be still in service. Battlefield reports recorded that the gun crews praised its reliability and mobility. What they did not like was the lack of protection. They complained that the armour was too thin and the superstructure was too tall. Its high profile made them an easy target to be spotted and then fired upon. There was poor protection at the rear of the superstructure. The loaders working at the back end of the vehicle were the most exposed to enemy fire. Another complaint was that the fighting compartment was too cramped.

The *Munitionsschlepper* Ammunition Carrier

There was not a lot of room in the Wespe for the crew to work the gun and carry the two-part 10.5-cm high-explosive shells. The *Munitionsschlepper auf Fahrgestell Panzerkampfwagen* II, or more simply '*Munitionsschlepper auf* Wespe', was designed as an unarmed armoured ammunition carrier.

It could carry ninety storage boxes containing 10.5-cm shells and cased cartridges filled with propellant. Records show that 159 were built in by the same Polish factories. Two would be attached to each *Abteilung* (battalion) of six Wespes. They were designed so that they could be converted in the field to a regular Wespe SPGs if necessary.

When there were not enough armoured *Munitionsschlepper* ammunition carriers available other forms of transport were used to keep the Wespe SPGs supplied with shells, such as the Opel Blitz lorry, half-tracks such as the Sd.Kfz 7 or 11.

Surviving SPGs

Deutsches Panzermuseum (German Tank Museum) Munster, Germany.
Musée des Blindés (French Tank Museum) Saumur, France (The only survivor with the original muzzle brake).
Kubinka Tank Museum, Russia.
Wespe Wreck, Battle of Normandy Museum, Bayeux, France.
Two Wespe Wrecks, André Becker Private Collection, Belgium

Specifications

Length: 4.81 m (15 feet 9 inches)
Width: 2.28 m (7 feet 6 inches)
Height: 2.30 m (7 feet 7 inches)
Total weight, battle ready: 11 tonnes (10.82 tons)
Crew: Five (driver, commander, gunner, and two loaders)
Propulsion: Maybach HL 62 TR six-cylinder, water-cooled 6.23-litre 140-hp gasoline/petrol engine
Speed (on/off road): 40/20 km/h (25/12.5 mph)
Maximum range (on/off road): 220/140 km (135/85 miles)
Main armament: 10.5-cm leFH 18/2 L/26 (early) or L/28 (late)
Secondary armament: 7.92-mm (0.31 inches) MG 34 or MG 40
Main gun elevation: -5 to +42 degrees
Upper hull armour: 15 mm (0.59 inches)
Lower hull armour: 18 mm (0.70 inches)
Side hull armour: 15 mm (0.59 inches)
Superstructure armour: 10 mm (0.39 inches)
Production total: 676

10.5-cm leFH 18/2 (Sf) *auf Fahrgestell Panzerkampfwagen* II Wespe (Sd.Kfz.124) artillery self-propelled gun from the 146th *Panzer Artillerie Regiment*, Panzer Lehr Regiment, Normandy, summer 1944. (*David Bocquelet*)

Above: 10.5-cm leFH 18/2 (Sf) *auf Fahrgestell Panzerkampfwagen* II Wespe (Sd.Kfz.124) artillery self-propelled gun from the 1st *Abteilung*, *Panzerartillerie Regiment*, 8th *Panzerdivision*, Ukraine, summer 1944. (*David Bocquelet*)

Below: 10.5-cm leFH 18/2 (Sf) *auf Fahrgestell Panzerkampfwagen* II Wespe (Sd.Kfz.124) artillery self-propelled gun from an unidentified unit, Italy, summer 1944. (*David Bocquelet*)

10.5-cm leFH 18/2 (Sf) *auf Fahrgestell Panzerkampfwagen* II Wespe (Sd.Kfz.124) artillery self-propelled gun from an unidentified *Abteilung*, perhaps part of the Hermann Göring Panzer Division, Anzio, 22 January 1944. (*David Bocquelet*)

Above: 10.5-cm leFH 18/2 (Sf) *auf Fahrgestell Panzerkampfwagen* II Wespe (Sd.Kfz.124) artillery self-propelled gun of an unidentified unit, Hungary, March 1945. (*David Bocquelet*)

Below: Munitionsschlepper auf Wespe, *Fallschrimpanzerdivision* Hermann Göring, East Prussia, winter 1944–45. (*David Bocquelet*)

12

15-cm sIG 33 (Sf) *auf Geschützwagen* 38(t) *Grille* (Sd.Kfz.138/1)

The German word *Grille* means 'cricket', a small insect that makes a very large noise. The German *Grille* self-propelled artillery gun was a small vehicle that made a very large noise when its 15-cm howitzer was fired. Two versions were made.

Its official designation was 15-cm *schweres Infanteriegeschütz* 33/1 *auf Selbstfahrlafette* 38(t) Ausf. H (Sd.Kfz. 138/1). It was based on the Czech-built Panzer 38(t) Ausf. H tank chassis that had the engine at the rear of the tank. A later version of the *Grille* used the Ausf. M Panzer 38(t) tank chassis that had the engine in the middle of the vehicle. The words Ausf. H refer to the model of Panzer 38(t) tank chassis that was used. The letter 't' after the number '38' indicates that the vehicle is of Czechoslovakian origin. The vehicle was also shortened to 15-cm sIG 33 (Sf) *auf Geschützwagen* 38(t). The word *Geschützwagen* translates to 'gun vehicle', often shortened to GW. You can see why just referring to it by its nickname, *Grille*, was easier.

With the decision made by the German High Command to invade the Soviet Union in 1941, the need for artillery guns on tracked chassis that could keep up with the panzer divisions, became more important. After the occupation of Czechoslovakia, the Skoda factory was used to produce the Panzer 38(t) tank. By 1941, this tank was considered obsolete for front-line duties. The turret was too small to take a more powerful anti-tank gun but the chassis and engine were tried and tested. It was a reliable vehicle.

Design and Production

Initially, Skoda engineers were given the task of developing a self-propelled artillery gun (SPG) that carried the 15-cm sIG 33 howitzer, a five-man crew and weighed around 16 tonnes. Design work started in July 1941 but the project was taken away from them on 1 September 1942 and reassigned to the makers of the 15-cm sIG 33 howitzer, Rheinmetall-Borsig. The reason for this is not known: there could have been technical problems with mounting the gun or because the design team at Skoda were taking too long to produce a prototype.

The gun manufacture's engineers designed a new gun mount and called the gun with this configuration the 15-cm sIG 33/1. Once fitted to the stripped-down Panzer 38(t) Ausf.

M chassis they started work on fitting a protective armoured open-topped superstructure casement to protect the crew from small arms fire and high-explosive shell fragments.

The front armour of the casement was 25 mm thick and the sides were 14 mm thick. In harsh weather, a canvas tarpaulin was tied down over the open top. It was also used as a sun shield in very hot weather. A H-shaped gun lock was fitted to the front of the hull to support the heavy gun barrel when the vehicle was being driven across rough ground.

It was powered by the same engine used in the Panzer 38(t) tank: the Praga TNHPS/II six-cylinder, water-cooled, 7.75-litre, 150-hp petrol/gasoline engine. The transmission was a Praga/Wilson CV/THNP, which had five forward and one reverse gear.

In the factory, it was known as the *Geschützwagen* 38 *für* 15-cm sIG 33/1 (Sf). The first prototype was completed in late 1942 and successfully completed trials. The German Army ordered 200. The Böhmisch-Mährische Maschinenfabrik (BMM) Company based in Prague was instructed to complete the work. It used to be called Českomoravská Kolben-Daněk (ČKD Praga) and had been involved in manufacturing the LT vz.34, LT vz.35, Panzer 35(t), and Panzer 38(t) tanks. In October 1943, a further ten 15-cm sIG 33/1 (Sf) *Grille* SPGs were ordered.

Operational Service

Self-propelled heavy infantry gun companies were issued six *Grille* sIG 33/1 SPGs each. They were attached to *Panzer-Grenadier-Regiments*: some regiments had more than one company. In July 1943, SS-Division 'Das Reich' had two *Grille*-equipped companies during Operation *Zitadelle*, the Battle of Kursk. At the same battle, the *Panzer-Grenadier-Regiment Großdeutschland* (Pz.Gr.Div. 'G.W.') fielded sixteen *Grille* sIG 33/1 SPGs.

The 2nd *Grille* Artillery SPG

The Panzer 38(t) chassis was redesigned with the engine moved to the middle of the vehicle by the team developing the *Panzerjäger* 38 tank hunter SPGs. It was called the Panzer 38(t) Ausf. M.

Moving the engine to the middle gave better weight distribution on the suspension and more room for the crew to work the gun at the rear of the vehicle. The vehicle was fitted with a more powerful Praga AC six-cylinder, water-cooled, 7.75-litre, petrol/gasoline 150-hp engine.

Design and Production

The BMM factory was awarded the contract to build 200 of these new 15-cm sIG 33/2 (Sf) *auf Geschützwagen* 38(t) *Grille* (Sd.Kfz.138/1) self-propelled artillery guns. The 15-cm sIG 33/2 heavy infantry gun was now mounted behind the engine. The gun crew were protected from small arms fire and high-explosive shell shrapnel by a smaller open-topped superstructure casement that only covered the rear half of the vehicle. The front and side armour of this protective shield was 10 mm thick.

The same Praga/Wilson CV/TNHP transmission was used. It had five forward and one reverse gear. In the factory, it was known as the *Geschützwagen* 38 M *für* 15-cm sIG 33/2 (Sf). In December 1943, the first fourteen 15-cm sIG 33/2 (Sf) *auf* GW 38(t) SPGs drove out of the BMM factory doors. By August 1944, a further 138 had been produced.

The company were instructed to give priority to building the *Jagdpanzer* 38 Hetzer SPG. Production stopped in September 1944. A further seventeen were built when work resumed in February and March 1945.

Operational Service

The new Ausf. M chassis *Grille* SPGs were sent as replacement vehicles to units that had lost their Ausf. H versions through battle damage or capture. It was common to see sIG Kp.(Sf) heavy infantry self-propelled gun companies operating both versions of the *Grille* at the same time.

Surviving SPGs

Fahrzeugmuseum, Bad Ischl, Austria
US Army Artillery Museum, Fort Sill, OK, USA
Munitionspanzer 38(t): US Army Artillery Museum, Fort Sill, OK, USA

Specifications

Length: 4.95 m (16 feet 3 inches)
Width: 2.15 m (7 feet 1 inch)
Height: 4.95 m (16 feet 3 inches)
Total weight, battle ready: 11.5 tonnes (11.31 tons)
Armament: 15-cm (5.9 inches) sIG 33 howitzer, fifteen rounds
Armour: 10–50 mm (0.39–1.96 inches)
Crew: Four (driver, commander, gunner, and loader)
Propulsion: Praga TNHPS six-cylinder, water-cooled, 7.75-l, petrol 150-hp engine
Speed: 35 km/h (22 mph)
Suspension: Torsion bar
Range: 190 km (120 miles)
Production: 282

Above: 15-cm sIG 33 (Sf) *auf Geschützwagen* 38(t) Ausf. H *Grille* (Sd.Kfz.138/1) artillery self-propelled gun of the 9th *Kompanie*, 113th Panzergrenadier Regiment, Russia, 1943. (*David Bocquelet*)

Below: 15-cm sIG 33 (Sf) *auf Geschützwagen* 38(t) Ausf. H *Grille* (Sd.Kfz.138/1) artillery self-propelled gun of the 9th *Kompanie*, Panzergrenadier Regiment 2, 2nd Panzerdivision, Normandy, summer 1944. (*David Bocquelet*)

15-cm sIG 33 (Sf) *auf Geschützwagen* 38(t) Ausf. H *Grille* (Sd.Kfz.138/1) artillery self-propelled gun of the *Panzergrenadier* Regiment 9/67 or 26, 26th *Panzerdivision*, Italy, 1944. (*David Bocquelet*)

Above: 15-cm sIG 33 (Sf) *auf Geschützwagen* 38(t) Ausf. H *Grille* (Sd.Kfz.138/1) artillery self-propelled gun of the *Panzergrenadier* Regiment 901, Russia, 1944. (*David Bocquelet*)

Below: 15-cm sIG 33/1 (Sf) *auf Geschützwagen* 38(t) Ausf. K *Grille* (Sd.Kfz.138/1) artillery self-propelled gun in winter whitewash livery. (*David Bocquelet*)

15-cm sIG 33/1 (Sf) *auf Geschützwagen* 38(t) Ausf. K *Grille* (Sd.Kfz.138/1) artillery self-propelled gun of an unknown unit in Russia 1944. (*David Bocquelet*)

Above: 15-cm sIG 33/1 (Sf) *auf Geschützwagen* 38(t) Ausf. K *Grille* (Sd.Kfz.138/1) artillery self-propelled gun in East Prussia 1945. (*David Bocquelet*)

Below: Munitionspanzer 38(t) Ausf.K 'Grille' (Sd.Kfz.138/1) artillery self-propelled gun in Germany May 1945. (*David Bocquelet*)

13

15-cm sFH 18/1 (Sf) *auf Geschützwagen* III/IV Hummel (Sd.Kfz.165)

The full designation of this self-propelled artillery gun (SPG) was *Panzerfeldhaubitze* sFH 18/1 (Sf) *auf Geschützwagen* III/IV Hummel, Sd.Kfz. 165. The German word '*Hummel*' means bumblebee. This armoured fighting vehicle had a nasty sting. A vehicle fitted with an artillery field howitzer was called a *Geschützwagen*. In February 1944, an official order was received to stop using the nickname 'Hummel'. The German High Command did not think it correct to call a large self-propelled artillery gun after a little flying insect.

Design and Production

A total of 705 Hummel 15-cm self-propelled artillery guns were built between February 1943 and March 1945. Some 157 Hummel *Munitionsträger* ammunition carriers were also produced. The Hummel was designed in 1942. The prototype featured a very large muzzle brake, but this was not used on the production models. The contract was awarded to the Alkett company. Deutsche Eisenwerke in Duisburg was contracted as the assembly firm. The first five production series Hummels were completed in February 1943 and entered service in March 1943. They were sent immediately to the Eastern Front to provide artillery support for the panzer divisions.

The initial contract for 500 (including the ammunition carrying *Munitionsträger* version) was completed in January 1944. The new improved version of the Hummel appeared in early 1944. Due to Allied bombing, the Hummel and the parts needed for its construction were built at a number of different factories: Deutsche Edelstahlwerk AG, Krefeld; Deutsche Röhrenwerke AG, Werk Thyssen, Mülheim an der Ruhr; Spreewerk and M.A.N.; Nordbau; Deutsche Edelstahlwerk AG, Werk Stahlindustrie in Duisburg; and Deutsche Edelstahlwerk AG, Werk Teplitz-Schonau.

The powerful 15-cm sFH 18 L/30 heavy field howitzer was mounted on a specially designed Alkett/Rheinmetall-Borsig lengthened German tank chassis called the *Geschützwagen* III/IV. Components were adopted from both the Panzer III and the Panzer IV tank chassis.

The more robust final drive wheels, front drive wheels and steering units plus the Zahnradfabrik SSG 77 transmission gearbox were adopted from the Panzer III Ausf. J. The Maybach HL 120 TRM engine with its cooling system, the suspension, and idler with track tension adjustment were adopted from the Panzer IV.

The engine was moved from the rear of the tank chassis to the centre of the vehicle to make room for the gun and the armoured fighting compartment at the back of the SPG. The *Geschützwagen* III/IV hull was also used for mounting the 88-cm anti-tank gun. This self-propelled gun was called the Nashorn.

Unlike the Nashorn's armour-piercing rounds, the Hummel's 15-cm high-explosive shells came in two parts. The explosive shell was loaded first, followed by the variable charge canister. This meant that the Hummel could only carry eighteen rounds of HE.

When not in use the Hummel's 15-cm howitzer was locked in place by a large 'A' frame travel-lock bracket that was mounted on the front hull glacis-armoured plate. This stopped the gun moving up and down too violently when the vehicle was travelling across the rough undulating ground.

On early versions of the Hummel, the front top of the hull had sloping armour with a raised armoured compartment for the driver on the left of the vehicle. The front hull superstructure and driver's armoured compartment were redesigned in early 1944 and enlarged, covering the whole width of the vehicle. The radio operator and driver now had more space to work in.

The exhaust system was also changed on the later model. It was moved from the original location below the rear double doors. The exhaust mufflers were dropped and the end of the exhaust pipes were cut at a slant away from the tracts to avoid stirring up any additional dust.

The *Geschützwagen* III/IV tank chassis did not have a hull-mounted machine gun. Crews were issued with a single 7.92-mm MG 34 recoil-operated air-cooled machine gun (or later a MG 42), carried inside the fighting compartment, and two 9-mm *Maschinenpistole* (MP) 38 submachine guns for self-defence.

The Hummel was designed to be operated by a crew of six: commander, driver, and four gunners. They were protected by an enclosed high silhouette armoured fighting compartment. Although it was open-topped, the crew were issued with a thick canvas tarpaulin covers that could be used in bad weather.

In front of the driver, a metal wire grid was fixed into position to aid the driver manoeuvring the vehicle in the correct fire position. Some early versions of the Hummel had a metal pole and wire mesh roof framework fitted above the fighting compartment of the vehicle. These were designed to prevent grenades and mines being thrown into the vehicle as it moved through towns and cities.

The Hummel SPGs were fitted with *Funksprechgerät* f wireless radio sets (Fu.Spr. Ger.f) for communication with other vehicles and the battery commander. For internal communications, the crew used a different system called the *Bordsprechgerät*. The driver was a long way forward from the fighting compartment and the engine and tracks made a lot of noise when the vehicle was on the move. This is why an internal communication system was fitted.

A metal-louvred cover ventilated the engine and radiator. They were on the side of the superstructure. It was found that they often were caked in dirt and dust which reduced the efficiency of their cooling function. Later versions were fitted with an angled shield that opened upward protecting the vents from clogging up. Air was drawn down from the top. These cooling air intake and exhaust covers were fitted to Hummels around August 1943.

Three red-and-white-striped aiming stake poles were carried in brackets below the rear door. The gunner would use a large ZE 34 sight. The top lens aperture would point to the rear of the vehicle. The gunner uses this aperture of the sight to locate the aiming sticks that a member of the crew had pounded into the ground at the rear at a known bearing from the vehicle by using a compass (compasses did not work inside a metal vehicle in 1943). By lining up the red and white fire-aiming stake, he would be able to work out the correct bearing the gun barrel is pointing towards.

The upper fighting compartment superstructure walls were constructed using 10-mm (0.39-inch) thick E11 chrome-silicon armour plates hardened to 153 kg/mm^2 for protection against shell fragments. The 30-mm (1.18-inch) thick front hull was made using face-hardened FA32 armour plates. The rest of the hull was made from cheaper rolled SM-Stahl (carbon steel) that was hardened to 75–90 kg/mm^2. It took 20-mm (0.78-inch) thick plates of SM-Stahl to provide equivalent protection against penetration by SmK (7.92-mm AP bullets) as 14.5 mm (0.57 inches) of E11 armour plate.

The early Hummel SPGs used the standard 1943 38-cm wide SK18 track that had three smooth metal pads visible on the front face of the track. In winter, some vehicles were fitted with track width extenders called *Winterketten* (winter track). These triangular pieces of metal were bolted on to the outer edge of the track to extend the width of the track and help the vehicle move across snow and mud by spreading the load over a larger area. They were problematic: they fractured and often fell off.

In 1944, vehicles started to be fitted with the wider *Ostketten* (east track) to cope with the conditions found on the Eastern Front. The *Winterketten* extensions made the SK18 tank track 55 cm wide. The one-piece *Ostketten* was 56 cm wide and did not have bits falling off it.

On the top edge of the superstructure, there are a number of D-shaped rings fitted to the armour plate. These were used to strap down the bad weather tarpaulin. It was also used in the hot continental summers to give shade from the sun.

At the rear of the vehicle, below the hatch doors, a towing bar was fitted. This enabled the vehicle to be towed out of the mud if it got stuck. It could also tow other vehicles, trailers, or supply sledges.

Operational Service

Hummels were initially used on the Eastern Front. The German Army Wehrmacht and SS panzer divisions each had their own heavy self-propelled artillery battery as part of their Artillery Regiment battalion. Each battery normally consisted of six Hummels supplied by one or two *Munitionsfahrzeuge* Hummel (Muni-Kfz) armoured ammunition carrier.

In March 1943, the first batch of eight Hummel SPGs entered service followed by another 46 in April. A few months later they saw their first action during Operation *Zitadelle* (Citadel), which started on 5 July 1943 on the Eastern Front. Ten divisions were issued with Hummels. A total of sixty were available to be deployed during this battle. They were used by *Panzer-Artillerie-Regiments* on the Eastern Front until the end of the war. A small number were captured by the Soviet Red Army and used against Axis forces in Hungary. Hummels were used in Greece, Italy, and north-west Europe in 1944.

The organisation of Artillery Regiments varied over time and location. The authorised establishment of the Panzer Artillery Regiment of the *Heer* Panzer Division that took part in the Battle of the Bulge, Ardennes Offensive in December 1944 had three *Abteilungen* (battalions). The second and third battalion comprised of towed 10.5-cm, 15-cm, and 17-cm howitzers but the first battalion was equipped with self-propelled artillery guns:

1.*Abteilung*
Stabskompanie (HQ company)
1st Battery (6 × Wespe 10.5-cm Artillery SPG)
2nd Battery (6 × Wespe 10.5-cm Artillery SPG)
3rd Battery (6 × Hummel 15-cm Artillery SPG)

The structure of the SS *Panzer-Artillerie-Regiment* 2 belonging to the 2nd SS Panzer Division 'Das Reich' in 1944 was different. It had five batteries in its artillery *Abteilung* (battalion):

1st Battery (12 × 17-cm K18 Howitzer)
2nd Battery (6 × 15-cm Hummel SPG)
3rd Battery (12 × 15-cm Howitzer)
4th Battery (12 × 10.5-cm Nashorn SPG)
5th Battery (12 × 10.5-cm Howitzer)
Forty Motorcycle troops

The *Panzer-Artillerie-Regiment* 103 was posted south of Orel, a city 360 km south-west of Moscow, between the first week of July and the middle of August 1943. Its II. *Abteilung* (2nd Battalion) had three batteries. The commander reported that the 2:1 ratio of 10.5-cm and 15-cm artillery guns met all tactical requirements.

1st Battery (6 × Wespe 10.5-cm Artillery SPG)
2nd Battery (6 × Wespe 10.5-cm Artillery SPG)
3nd Battery (6 × 15-cm Hummel SPG)

Mechanical Problems

The Hummel was prone to some mechanical defects. These became evident after they had been used in action. The 15-cm heavy field gun caused stresses on the vehicle when it was fired. The recoil force was felt throughout the vehicle and some mechanical parts failed. Bolts, screws, and rivets worked loose; this caused equipment and fittings to come away from the hull as well as leaking. Gears failed in steering units. Fan drives snapped or came off.

When new targets were assigned to the battery, the vehicle would have to be turned on the spot to align the gun to the new map reference. Idler wheel mountings broke. The final drive large gear failed. Bolts that fixed the final drive snapped. The demand for spare parts for the Hummel sometimes could not be met.

Hummels were used in Italy. They had to negotiate the small winding mountainous roads. Continually swivelling on the spot to get around tight turns also caused the final drives to break down. Going down steep gradients wore out brake linings quickly. They had to be regularly replaced. Unavailability due to mechanical problems was much higher in Italy than on the Eastern Front due to the demands of the terrain. If these vehicles could not be fixed during a retreat, they had to be blown up to stop them falling into Allied hands.

Czechoslovakian Army Hummels

Surviving Hummel Artillery self-propelled guns were used by the Czechoslovakian Army after the Second World War. Twelve vehicles underwent renovation and entered service in 1950. They were officially called *Samohybné děla* Hummel (152-mm ShH vz. 18/47N, SD-152). They were later withdrawn from Army service and presumably scrapped.

Munitionsträger Hummel

The self-propelled artillery units that operated Hummels needed to be supplied with ammunition regularly. As each vehicle could only carry eighteen rounds, they soon depleted their stock of shells.

The gun crew working the Hummel's 15-cm howitzer were protected by the vehicle's armour plate from small arms fire and high-explosive shell shrapnel fragments. Soft-skinned lorries carrying ammunition near the front line were liable to explode in that hostile environment.

The artillery regiments of the *Wehrmacht* used standard production Hummels that were not fitted with a gun to carry ammunition. They were called *Munitionsträger* Hummel Sd.Kfz 165. They were also known as *Munitionsfahrzeuge* Hummel, *Geschützwagen* III/IV *für Munition* (Sd.Kfz.165), or G.W.III/IV *für Munition*. In official documents, the abbreviation Muni-Fzg was used. Some 157 armoured ammunition carriers based on the Hummel *Geschützwagen* III/IV hybrid tank chassis were constructed between May 1943 and November 1944.

A flat 10-mm armour plate was bolted to the fighting compartment front to replace the normal gun shield. It was bolted rather than welded in place so that it could be removed and a gun fitted if necessary. The cargo space inside the armoured compartment was 15 cubic meters (530 cubic feet). It was the weight of the load carried, not the number of shells and propulsion cartridges, that was the deciding issue on how much could be carried. If the *Munitionsträger* was overloaded, it would have difficulty crossing rough ground and its suspension could fail.

Ammunition was delivered to the working Hummel battery in wicker tube-shaped shipping containers, each containing one 42.9-kg (94.6-lb) high-explosive projectiles. Artillerymen referred to the containers as a *Koffer* (suitcase). The separate charge cartridges arrived in wooden boxes.

Production Changes and Identification

One of the easiest ways of telling the difference between an early and late production Hummel when studying a photograph is to look at the upper front superstructure on the tank chassis. If you can see a separate armoured raised box on the left of the vehicle, where the driver would sit, then this is an early production vehicle. If there is a raised box that covers the whole width of the vehicle than this is a late production version. This change was introduced on the production line around February 1944.

If you cannot see the front of the vehicle, look for a large exhaust muffler/silencer box under the two small hinged doors at the rear of the vehicle. If you see one, you are looking at an early production vehicle. It was dropped from the late production models and two replacement spare bogie wheel holders were put in its place as well as an angled metal foothold to help the crew climb into the back of the vehicle. The end of the twin exhaust pipes can now be seen over the top of the track at the rear, one on each side.

This was a clever modification as the exhaust gases blew down on to the dust kicked up by the tracks during the summer and helped reduce the amount of dust that entered the crew compartment. These changes were introduced around August 1943.

On new drawings dated 15 May 1943, the fitting of two Bosch headlights at the front of the vehicle was stopped. Later production Hummels were now only fitted with one Bosch headlight over the front left track.

It is easy to tell the difference between a 15-cm Hummel and 10.5-cm Nashorn self-propelled artillery gun even though they both used the same *Geschützwagen* III/IV tank chassis. The 15-cm howitzer used in the Hummel was not fitted with a muzzle brake on the end of the gun barrel. The Nashorn's 10.5-cm gun always had a muzzle brake fitted.

Surviving SPGs

Deutsches Panzermuseum (German Tank Museum) Munster, Germany
Musée des Blindés (French Tank Museum) Saumur, France
Technik Museum, Sinsheim, Germany
Artillerie Schule, Idar-Oberstein, Germany
The Fort Sill Field Artillery Museum, Oklahoma, USA

Specifications

Length: 7.17 m (23 feet 5 inches)
Width: 2.97 m (9 feet 7 inches)
Height: 2.81 m (9 feet 2 inches)
Total weight, battle ready: 22 tonnes (21.65 tons)
Crew: Six (commander, driver, and four *Kanoniere* gun crew)
Propulsion: Twelve-cylinder water-cooled Maybach HL 120 TRM 11.9-litre petrol engine, 265 hp at 2,600 rpm

Fuel capacity: 400 litres
Top road speed: 42 km/h (26 mph)
Operational range (road): 215 km (133 miles)
Main armament: 15-cm sFH 18/1 L/30 howitzer with eighteen rounds
Secondary armament: 7.92-mm MG 34 machine gun
Armour: front 30 mm, sides 20 mm, rear 20 mm.
Superstructure armour: front 10 mm, sides 10 mm
Total production: 705

Above: 15-cm sFH 18/1 (Sf) *auf Geschützwagen* III/IV Hummel (Sd.Kfz.165) artillery self-propelled gun prototype with large muzzle brake. (*David Bocquelet*)

Opposite above: Early version of the 15-cm sFH 18/1 (Sf) *auf Geschützwagen* III/IV Hummel (Sd.Kfz.165) artillery self-propelled gun in winter whitewash livery Eastern Front, 1943. (David Bocquelet)

Opposite middle: Early version of the 15-cm sFH 18/1 (Sf) *auf Geschützwagen* III/IV Hummel (Sd. Kfz.165) artillery self-propelled gun with wire mesh protective roof, summer 1943. (*David Bocquelet*)

Opposite below: Late production 15-cm sFH 18/1 (Sf) *auf Geschützwagen* III/IV Hummel (Sd. Kfz.165) artillery self-propelled gun of the 2 SS-Pz Div '*Das Reich*', summer 1944. (*David Bocquelet*)

Clausewitz

Late production 15-cm sFH 18/1 (Sf) *auf Geschützwagen* III/IV Hummel (Sd.Kfz.165) artillery self-propelled gun with armoured engine-louvred exhaust cover. Summer 1944. (*David Bocquelet*)

Late production *Munitionsträger* Hummel ammunition carrier, summer 1944. (*David Bocquelet*)

14

10.5-cm leFH 18/6 *auf Waffenträger Geschützwagen* III/IV *Heuschrecke* IVb Grasshopper

The German 10.5-cm leFH 18/6 *auf Waffenträger Geschützwagen* III/IV *Heuschrecke* IVb was designated a weapon carrier (*Waffenträger*) and not a self-propelled artillery gun. The reason for this is that the turret could be removed from the top of the modified panzer IV tank chassis by a block and tackle rig attached to a movable metal frame.

The idea was that the gun crew could keep up with the armoured panzer divisions. When needed to fire as an artillery battery, to give long-range support firing high-explosive shells over the heads of the German infantry and tank crews, the gun would be removed and placed on the ground where it could be fired like a normal artillery gun.

The heavy lifting metal framework could be swung upright into position by a hydraulic system or a manual back up system. When not needed, it was lowered down and stored on top of the upper track guards on both sides of the tank chassis.

The vehicle could carry eighty-seven high-explosive shells. If more were needed, the turret could be removed and placed on a gun carriage and towed behind the tank chassis. This allowed for more ammunition to be carried on to the battlefield. The modified Panzer IV tank chassis became a turretless armoured ammunition carrier. This configuration would have only worked in gently undulating countryside or on roads. The gun carriage wheels and frame were carried on the tank chassis at the rear.

The 10.5-cm howitzer could also be fired from on top of the tank chassis. There was no top to the turret. There were a few disadvantages of an open-topped vehicle. The crew was exposed to the elements and were also at risk of injury from enemy-thrown hand grenades, mortars, and shrapnel from air burst enemy shells. A canvas tarpaulin rain cover was produced.

Not A Movable Pillbox

Some books have argued that the reason for removing the turret was because it could be used as an armoured pillbox. This was not its function. It was an artillery gun that was designed to operate behind the front line. It was not an anti-tank gun.

The protective armour that surrounded the gun was not of a thickness that would have stopped armour-piercing tank shells. It was only designed to protect the gun crew from small arms fire as well as high-explosive shell and mortar round shrapnel fragments.

Two Competing Models

The German armament factory of Alkett and Rheinmetall-Borsig based near Berlin had come up with a similar prototype design called the 10.5-cm leFH 18/40/2 *auf Geschützwagen* III/IV. It did not have the lifting gear on the side of the vehicle but the turret was removable just like the Krupp-Gruson design.

It used the standard Panzer IV tank chassis and had slightly better performance than the Krupp-Gruson's *Heuschrecke* IVb Grasshopper. The Alkett Rheinmetall-Borsig model was completed in March 1944.

Design

In May 1943, German Army weapon designers decided to build a prototype *Heuschrecke* IVb. It would be built using a Hummel SPG chassis and a 10.5-cm LeFH 18/l light field howitzer in a removable turret.

In June 1943, the Krupp Grusonwerk factory started work on building this new armoured fighting vehicle using a new Hummel chassis number 320148. Other sources state that three prototypes were built, with serial numbers 582501, 582502, and 582503.

The Hummel self-propelled artillery gun had powerful 15-cm sFH 18 L/30 heavy field howitzer mounted on a specially designed Alkett/Rheinmetall-Borsig lengthened German tank chassis called the *Geschützwagen* III/IV. It was also referred to as the IVb.

These prototypes were referred to as the *Heuschrecke* 10 or *Heuschrecke* IVb. The German word '*Heuschrecke*' means Grasshopper. It was quite appropriate. The long-folded metal lifting equipment kept on top of each track mudguard looked like a grasshopper's insect legs. The number '10' refers to the size of the gun: the 10.5-cm. The number IVb refers to the modified Panzer III/IV tank chassis.

Components were adopted from both the Panzer III and Panzer IV tank chassis. The more robust final drive wheels, front drive wheels and steering units plus the Zahnradfabrik SSG 77 transmission gearbox were adopted from the Panzer III Ausf. J.

The Maybach HL 120 TRM engine with its cooling system, the suspension, and idler with track tension adjustment were adopted from the Panzer IV. The engine was moved from the rear of the tank to the centre of the vehicle to make room for the gun and the armoured fighting compartment at the back of the SPG.

The *Geschützwagen* III/IV tank chassis did not have a hull-mounted machine gun. Crews would be issued with a single MG 34 or MG 42 machine gun, carried inside the fighting compartment, for self-defence.

The Krupp-Gruson designers envisioned that the *Heuschrecke* IVb would start to replace the 10.5-cm leFH 18 *auf Fahrgestell Panzerkampfwagen* II Wespe self-propelled artillery gun in May 1944.

The tank engineers at the Krupp Grusonwerk armaments factory made changes to the superstructure and chassis to enable the *Heuschrecke* turret to be fitted and the installation of the hydraulic mechanism needed to dismount the turret.

The Hummel was powered by a Maybach HL 120 TRM engine that was fitted in the middle of the vehicle to allow more room for the gun crew to work the gun at the back of the vehicle. This was changed for the 10.5-cm leFH 18/1 L/28 *auf Waffenträger* IVb prototype. The engine and the radiators were moved to the rear of the chassis.

The *Heuschrecke* IVb prototype turret was armed with the 10.5-cm leFH 18/1 L/28 light field howitzer. The production models, however, were to have the newer, more powerful 10.5-cm leFH 43 L/28.

Weapon Trials

The German Army Weapons Agency (*Heereswaffenamt*) sent weapon testing inspectors from the *Gliederung Waffenamt Prüfwesen* (Wa Prüf 4) artillery section to examine the new artillery SPG. They subsequently submitted a report following their inspection visit on 28 September 1943.

On the positive side, they noted it used mature tested parts. It could be traversed through 360 degrees and fired at high elevations when dismounted. The design worked and had adequate space for the stowage of equipment and ammunition. It could carry eighty-seven 10.5-cm shells. On the negative side, they concluded the 10.5-cm leFH 18/1 L/28 *auf Waffenträger* IVb would be expensive to produce and the dismounted turret was not mobile.

The first trials happened on 11 October 1943 at Hillersleben. The hydraulic arms were used to dismount the turret. It was found to be too heavy. A lighter redesigned turret was manufactured and ready for testing by the end of December 1943.

At the end of January 1943, to complement the hydraulic turret dismounting system, the design team at Krupp started work on a backup hand-powered system in case of problems with the hydraulics on the battlefield.

On 28 March 1944, the Wa Prüf 4 artillery weapon testing inspectors were present at a second demonstration of the modified 10.5-cm leFH 18/1 L/28 *auf Waffenträger* IVb.

Their recommendations after that visit were that a hand-operated crane, for dismounting the turret, be fabricated. Wheels were to be added to the dismounted turret frame and a standard gun carriage and recoil management recuperator cylinder from the leFH 18/40 gun was to be installed.

On 31 May 1944, the newly modified 10.5-cm leFH 18/1 L/28 *auf Waffenträger* IVb, with a parallelogram hand-operated crane and wheels for the dismounted carriage, were demonstrated to the Wa Prüf 4 artillery weapon testing inspectors.

This time their report conclusion stopped any further development and design work on this project. They concluded that the 3.8-tonne dismounted turret was unusable on the battlefield. The 10.5-cm leFH 18/1 L/28 *auf Waffenträger* IVb never entered mass production.

There was no dramatic advantage to building this weapon over the 15-cm Hummel, 15-cm Wespe, or 105-cm *Grille* self-propelled artillery guns already in production. These vehicles were less complicated to produce and operate.

Surviving SPG Prototype

When the American Army occupied Germany at the end of the war, they found a surviving 10.5-cm leFH 18/1 L/28 *auf Waffenträger* IVb prototype. It was shipped back to the US Army Ordnance Corps proving grounds at Aberdeen, Maryland for testing and evaluation. It was transferred to Fort Still in 2012 and the Grasshopper 10 was restored by the Fort Sill Directorate of Logistics paint shop.

Specifications

Length: 6.57 m (21 feet 7 inches)
Width: 2.9 m (9 feet 6 inches)
Height: 2.65 m (8 feet 3 inches)
Total weight, battle ready: 24 tonnes (23.62 tons)
Crew: Five (commander, driver, gunner, and two loaders)
Propulsion: Maybach HL 120TRM twelve-cylinder, water-cooled, 285-hp gasoline/petrol engine
Fuel capacity: 360 litres
Top road speed: 38 km/h (24 mph)
Operational range (road): 225 km (140 miles)
Main Armament: 10.5-cm leFH 18/6 howitzer with eighty-seven rounds
Secondary Armament: Handheld 9-mm machine pistol
Front hull armour: 30 mm
Sides and rear hull armour: 16–20 mm
Front turret Armour: 30 mm
Sides and rear turret armour: 15 mm
Total built: One or three

Factory prototype 10.5-cm leFH 18/6 *auf Waffenträger Geschützwagen* III/IV *Heuschrecke* IVb Grasshopper in *dunkelgelb* livery. (*David Bocquelet*)

10.5-cm leFH 18/6 *auf Waffenträger Geschützwagen* III/IV *Heuschrecke* IVb Grasshopper in grey livery. (*David Bocquelet*)

15

10.5-cm leFH 18/40/2 L/28 (Sf) *auf Geschützwagen* III/IV

The German weapons manufacturers Alkett Rheinmetall-Borsig tried to compete with the Krupp-Gruson to produce an armoured fighting vehicle that would carry a 10.5-cm *Leichte Feldhaubitze* 18/40/2 L/28 light field howitzer on to the battlefield like a self-propelled gun but then allow the gun to be lowered to the ground. The gun could also be fired from an armoured open-topped turret on top of the vehicle.

This then enabled the tank chassis to function as an armoured ammunition carrier. The driver could return to the nearest supply point and load the vehicle with more high-explosive shells and propulsion canisters.

The Krupp-Gruson design was called the 10.5-cm leFH 18/1 L/28 *auf Waffenträger* IVb. The Alkett Rheinmetall-Borsig design was given the name 10.5-cm leFH 18/40/2 L/28 (Sf) *auf Geschützwagen* III/IV.

To remove the turret and gun on the Krupp-Gruson design, a large folding metal gantry was mounted on the rear of the vehicle. When it was folded down the long metal arms were stowed on top of the track guards. This vehicle was given the nickname of the *Heuschrecke* (the Grasshopper) because of the way it looked: the folded metal arms looked like the legs of a grasshopper.

It is easy to tell the two designs apart. The Alkett Rheinmetall-Borsig 10.5-cm weapon carrier does not have any external folding metal struts on top of its track guards, unlike the Krupp-Gruson design.

The 10.5-cm LeFH 18/40/2 L/28 light field howitzer's carriage rails and wheels were stowed at the rear of the vehicle on the outside. When the gun crew wanted to dismount the howitzer from the vehicle to use it as a normal artillery gun on the floor, they would unlock the gun from its mount and use a jack to lift it up.

The armoured superstructure of the turret is hinged at the front to be folded forward to give more room to the gun crew during the removal procedure. The carriage, carriage wheels, and rear trails would then be reattached. A ramp down from the top of the vehicle would be fitted and a manual block and tackle winch system used to lower the gun down to the ground. The reverse procedure would be used to remount the gun on top of the vehicle.

The howitzer could be fired from on top of the vehicle like a normal self-propelled gun. It could traverse through 360 degrees as it was mounted on a turntable gun platform that was countersunk into the top of the tank chassis. There was no top to the turret.

Design and Testing

Alkett submitted its designs for a 10.5-cm howitzer weapons carrier self-propelled gun to the German Army Weapons Agency (*Heereswaffenamt*) *Gliederung Waffenamt Prüfwesen* (Wa Prüf 4) artillery section on 27 September 1943.

The first prototype was ready for testing on 28 March 1944. It was demonstrated to the Wa Prüf 4 weapons inspectors at Hillersleben. The gun crew took around fifteen minutes to dismount the howitzer and mount it on the newly designed *Schiesspilz* (firing pedestal). The gun could not be traversed a full 360 degrees on the floor.

The inspectors submitted a report that recommended the protective turret armour was to be increased in thickness, that the *Schiesspilz* be carried on the vehicle, and that the howitzer be modified that it could fire to the rear of the vehicle at an angle of -5 degrees.

Alkett suggested that the turret superstructure armour could be angled to increase the gun crew's protection. This suggestion was rejected as it decreased the amount of ammunition that could be stored.

The Wa Prüf 4 weapons inspectors re-evaluate the new modified prototype after a second demonstration on 28 May 1944. They were happy with the changes. As a Panzer III/IV self-propelled gun tank chassis was going to be used, it was felt that field trials would not be necessary as this chassis was already being used in the production of the Hummel and Nashorn self-propelled guns.

The design team at Alkett/Rheinmetall-Borsig had already lengthened a German tank chassis. It was called the *Geschützwagen* III/IV. Components were adopted from both the Panzer III and Panzer IV tank chassis. The more robust final drive wheels, front drive wheels and steering units plus the Zahnradfabrik SSG 77 transmission gearbox were adopted from the Panzer III Ausf. J. The Maybach HL 120 TRM engine with its cooling system, the suspension, and idler with track tension adjustment were adopted from the Panzer IV.

The first twenty-five production 10.5-cm leFH 18/40/2 L/28 (Sf) *auf Geschützwagen* III/IV were due to be completed by October 1944 but there were long delays at the factory. The German Army Weapons Agency (*Heereswaffenamt*) cancelled the order on 12 December 1944 before any were completed.

A revised design which did not feature the countersunk turntable gun platform was put forward and accepted. The gun would now only have a limited traverse when mounted on the tank chassis like a normal self-propelled gun. This was hoped to make production cheaper and simpler. An initial order for 250 of these vehicles was approved. The first thirty-five would be built by February 1945 and the rest to follow. There are no production report documents that suggest any of these vehicles were produced.

There was no dramatic advantage to building this weapon over the 15-cm Hummel, 15-cm Wespe, or 10.5-cm *Grille* self-propelled artillery guns that were already in production. These vehicles were less complicated to produce and operate.

Surviving SPGs

A surviving prototype is stored at the Kevin Wheatcroft Collection, England.

Specifications

Length: 7.19 m (23 feet 6 inches)
Width: 3 m (9 feet 8 inches)
Height: 2.87 m (9 feet 4 inches)
Total weight, battle ready: 25 tonnes (24.60 tons)
Crew: Five (commander, driver, gunner, and two loaders)
Propulsion: Maybach HL 120 TRM twelve-cylinder, water-cooled, 265-hp gasoline/petrol engine
Fuel capacity: 500 litres
Top road speed: 42 km/h (26 mph)
Operational range (road): 190 km (118 miles)
Main armament: 10.5-cm leFH 18/40/2 L/28 howitzer with eighty-five rounds
Secondary armament: Handheld 9-mm machine pistol
Front hull armour: 30 mm
Sides and rear hull armour: 20 mm
Turret armour: 10 mm
Total built: One

10.5-cm leFH 18/40/2 L/28 (Sf) *auf Geschützwagen* III/IV artillery self-propelled gun prototype. (*David Bocquelet*)

16

10.5-cm leFH-18/4 (Sf) *auf Geschützwagen* Lorraine Schlepper(f)

After the surrender of France in 1940, a lot of French Army military equipment was taken into operational use by the Germans. Some of the French tanks and armoured tractors like the Lorraine 37L were converted into self-propelled guns.

These vehicles would be able to keep up with the panzer divisions. There were two main types of self-propelled guns in the German Army during the Second World War. One was fitted with an anti-tank gun and the other with an artillery howitzer, like the 10.5-cm leFH-18/40 *auf Geschützwagen* Lorraine Schlepper(f) self-propelled artillery gun. The word '*Schlepper*' means tractor. The letter (f) indicates that the SPG's chassis was French.

Alkett, based near Berlin, converted twelve Lorraine 37L tractors into the 10.5-cm leFH-18/40 *auf Geschützwagen* Lorraine Schlepper(f) self-propelled artillery guns in 1942. They were a large manufacturer of armoured fighting vehicles for the German Army during the Second World War. German Army Major Alfred Becker worked with the company to create these conversions. The following year he was in Normandy at the head of a *Baukommando*, a construction command unit. Becker's men, engineers, and mechanics converted a further twelve Lorraine 37L tractors into self-propelled artillery gun by fixing 10.5-cm leFH-18/40 howitzers on to the top of these vehicles and making an armoured open-crew compartment.

The Lorraine 37L

During the First World War, the French and British Army needed a way to transport ammunition and supplies to the front line. Men and horses were getting killed and injured from small arms fire and exploding shell fragments. Tracked armoured supply vehicles were developed. This vehicle was developed by the Lorraine Company in 1937 as a replacement for the smaller Renault UE. It could transport a heavier load and was faster than the Renault UE. Production began in January 1939. By the time France surrendered in 1940, a total of 432 Lorraine 37L armoured supply tractors had been produced.

Operational Service

On D-Day (6 June 1944), the three battalions of the *Panzer-Artillerie-Regiment* 155 of the German Army's 21st *Panzerdivision* were equipped with a total of twenty-four 10.5-cm leFH-18/40 *auf Geschützwagen* Lorraine Schlepper(f) and twelve 15-cm sFH *auf Geschützwagen* Lorraine Schlepper(f) self-propelled artillery gun conversions.

Although suffering losses of tanks and assault guns, the official regimental records note that, four weeks after D-Day, none of the Lorraine conversions had been lost. That was not to last. Following the heavy fighting around Caen, the last ones were knocked out when they were caught in the Falaise Pocket and subjected to intense bombing, shelling, and gunfire in August 1944.

Identification

One of the easiest ways of telling the difference between an early 1942-built Alkett 10.5-cm leFH-18/40 *auf Geschützwagen* Lorraine Schlepper(f) and one built later by *Baukommando* Becker is to look at the horizontal angle joint on the side armour of the crew compartment. It was far lower on the earlier Alkett version.

21st *Panzerdivision*, June 1944

Panzer-Artillerie-Regiment 155
- I. *Abteilung* (1st Battalion)
  - 1st Battery 4 × 12.2-cm *Kanone* 390/1(r) (captured Soviet guns)
  - 2nd Battery 4 × 12.2-cm *Kanone* 390/1(r) (captured Soviet guns)
  - 3rd Battery 4 × 10-cm K 18s
- II. *Abteilung* (Sf) (2nd Battalion)
  - 4th Battery 6 × 10.5-cm leFH 18 *auf* Lorraine
  - 5th Battery 6 × 10.5-cm leFH 18 *auf* Lorraine
  - 6th Battery 6 × 15-cm sFH 13 *auf* Lorraine
- III. *Abteilung* (Sf) (3rd Battalion)
  - 7th Battery 6 × 10.5-cm leFH 18 *auf* Lorraine
  - 8th Battery 6 × 10.5-cm leFH 18 *auf* Lorraine
  - 9th Battery 6 × 15-cm sFH 13 *auf* Lorraine
- Rocket Battalion (*Werfer*)
  - 19th Battery 2 × S307(f) *R-Vielfachwerfer*

Surviving SPGs

Central Museum of the Second World War, Moscow, Russia
Private collection, France

Specifications

Length: 5.31 m (17 feet 5 inches)
Width: 1.57 m (5 feet 2 inches)
Height: 2 m (6 feet 7 inches)
Total weight, battle ready: 7.8 tonnes (7.7 tons)
Crew: Four (commander, driver, gunner, and loader)
Propulsion: Type 135 Delahaye six-cylinder 70-hp inline gasoline/petrol engine
Fuel capacity: 144 litres
Top road speed: 35 km/h (22 mph)
Operational range (road): 137 km (85 miles)
Main armament: 10.5-cm leFH 18 howitzer with twenty rounds
Front hull armour: 12 mm
Sides and rear hull armour: 9 mm
Superstructure armour: 8–14.5 mm
Total built: 166

10.5-cm leFH-18/40 *auf Geschützwagen* Lorraine Schlepper(f) (*Baukommando* Becker version) artillery self-propelled gun in *dunkelgelb* and green livery. (*David Bocquelet*)

Above: 10.5-cm leFH-18/40 *auf Geschützwagen* Lorraine Schlepper(f) (*Baukommando* Becker version) artillery self-propelled gun in *dunkelgelb* livery. 155th Panzer Artillery Regiment, 21st *Panzerdivision*, Normandy, summer 1944. (*David Bocquelet*)

Opposite above: 10.5-cm leFH-18/40 *auf Geschützwagen* Lorraine Schlepper(f) (Alkett Version) artillery self-propelled gun in *dunkelgelb* livery. 155th Panzer Artillery Regiment, 21st *Panzerdivision*, Normandy, summer 1944. (*David Bocquelet*)

Opposite middle: 10.5-cm leFH-18/40 *auf Geschützwagen* Lorraine Schlepper(f) (Alkett Version) artillery self-propelled gun in *dunkelgelb*, green, and brown livery. 155th Panzer Artillery Regiment, 21st *Panzerdivision*, Normandy, summer 1944. (*David Bocquelet*)

Opposite below: 10.5-cm leFH-18/40 *auf Geschützwagen* Lorraine Schlepper(f) (*Baukommando* Becker version) artillery self-propelled gun in grey livery. 155th Panzer Artillery Regiment, 21st *Panzerdivision*, Normandy, summer 1944. (*David Bocquelet*)

Below: 10.5-cm leFH-18/40 *auf Geschützwagen* Lorraine Schlepper(f) (*Baukommando* Becker version) artillery self-propelled gun in *dunkelgelb* and brown livery. 155th Panzer Artillery Regiment, 21st *Panzerdivision*, Normandy, summer 1944. (*David Bocquelet*)

17
10.5-cm leFH 18 (Sf) *auf Geschützwagen* 39H(f)

Putting a 10.5-cm howitzer on top of captured French Army Hotchkiss H39 Tanks chassis was a more efficient use of manpower from the traditional form of German artillery battery transportation. Two different howitzers were used during the development of this self-propelled artillery gun: the 10.5-cm leFH 16 (a First World War-era weapon) and the 10.5-cm leFH 18 gun.

Rain covers were produced. They covered both the crew compartment and the gun. The canvas was attached to the upper protective armour using the small D-shaped rings welded to the upper part of the structure.

As the French Hotchkiss H39 tank chassis was small, there was limited space for the storage of ammunition. Only thirty-six 10.5-cm HE two-part shells could be carried. The propellant charges were kept on the left of the vehicle while the projectile shells were stored on the right.

Shells were carried in the crew compartment via the rear of the vehicle. There were two large hinged armoured doors with a small protruding metal step ladder underneath them at the back.

When the ammunition supply lorries reached each battery, they would unload two different types of packing boxes. The propellant charges, that looked like giant tin cans, were transported in protective wooden-hinged boxes. The front projectile part of the shell was wrapped in its own individual open wooden crate.

A 7.92-mm MG 34 machine gun was attached to top right side of the armour casement, on a swivel mount. Spare ammunition (fifty-round drums) were stored underneath the mount. It fired 7.92-mm (0.31-inch) bullets.

Identification

One of the easiest ways of telling the difference between a 10.5-cm leFH 18 (Sf) *auf Geschützwagen* 39H(f) self-propelled gun and a 10.5-cm leFH 16 (Sf) *auf Geschützwagen* 39H(f) SPG is to look at the armoured housing that surrounds the gun's recuperator mechanisms. On the 10.5-cm leFH 18, there are two: one above and below the gun. On the First World War 10.5-cm leFH 16, there is only one below the gun barrel.

The 7.5-cm PaK 40/1 (Sf) *auf Geschützwagen* 39H(f) anti-tank self-propelled gun version of the Marder I is very similar in looks to the 10.5-cm leFH 16 (Sf) *auf Geschützwagen* 39H(f) self-propelled artillery gun. They both have a single armoured housing box surrounding the gun's recuperator mechanisms which is below the gun barrel.

The 7.5-cm PaK 40/1 (Sf) *auf* GW-39H(f) has a long piece of armour at the bottom of the gun's *Schutzschild* (splinter shield) that slopes at an angle towards the armoured casement below the gun. This is missing on the 10.5-cm leFH 16 (Sf) *auf* GW-39H(f).

Operational Service

At least twenty-four 10.5-cm LeFH 16 and 10.5-cm leFH 18 (Sf) *auf Geschützwagen* 39H(f) self-propelled guns went into action in Normandy in June 1944. They were part of the German Army assault gun battalion *Sturmgeschütz-Abteilung* 200 of 21. *Panzerdivision.*

The last ones were knocked out when they were caught in the Falaise Pocket and subjected to intense bombing, shelling, and gunfire in August 1944.

Each of the four batteries of *Sturmgeschütz-Abteilung* 200 had four 7.5-cm PaK 40 (Sf) *auf* GW 39H(f) each. While I. *Batterie* had six 10.5-cm leFH 16 (Sf) *auf* GW 39H(f), the other three batteries (II. *Batterie*, III. *Batterie*, and IV. *Batterie*) each had six 10.5-cm leFH 18 (Sf) *auf* GW 39H(f).

Surviving SPGs

Association de Sauvegarde du Patrimoine Historique Militaire, 1 Rue Georges Cuvier, La Watzenau, France
Musée des Blindés (French Tank Museum) Saumur, France
MM Park Military Museum near Strasbourg, France

Specifications

Length: 4.56 m (without gun) (14 feet 11 inches)
Width: 1.55 m (5 feet 1 inch)
Height: 2.02 m (6 feet 7 inches)
Total weight, battle ready: 13 tonnes (12.79 tons)
Crew: Five (commander, driver, gunner, and two loaders)
Propulsion: Hotchkiss six-cylinder, water-cooled, 5.97-litre petrol engine, 120 hp at 2,800 rpm
Fuel capacity: 207 litres
Top speed: 36 km/h (22 mph)
Operational range (road): 180 km (111 miles)
Main armament: 10.5-cm (4.13-inch) leFH 18 howitzer with thirty-six rounds
Secondary armament: 7.92-mm (0.31-inch) MG 34 machine gun
Armour: 22–34 mm (0.86–1.34 inches)
Total production: Forty-eight

10.5-cm leFH 16 (Sf) *auf Geschützwagen* 39H(f) artillery self-propelled gun in *dunkelgelb* livery and net for hanging branches of trees and bushes for camouflage. *Sturmgeschütz-Abteilung* 200, 21st *Panzerdivision*, Normandy, summer 1944. (*David Bocquelet*)

Above: 10.5-cm leFH 18 (Sf) *auf Geschützwagen* 39H(f) artillery self-propelled gun in dunkelgelb livery. *Sturmgeschütz-Abteilung* 200, 21st *Panzerdivision*, Normandy, summer 1944. (*David Bocquelet*)

Below: 10.5-cm leFH 18 (Sf) *auf Geschützwagen* 39H(f) artillery self-propelled gun in dunkelgelb and green livery, *Sturmgeschütz-Abteilung* 200, 21st *Panzerdivision*, Normandy, summer 1944. (*David Bocquelet*)

18

15-cm sIG 33/2 (Sf) *auf Jagdpanzer* 38(t)

The German Army-run factories in Czechoslovakia could produce *Jagdpanzer* 38(t) Hetzer tank hunter chassis faster than they could produce the main 75-mm (2.95-inch) gun. There were stocks of old 15-cm (5.9-inch) sIG 33/2 heavy infantry guns available to be mounted in those chassis.

Troops on the Eastern Front always needed more heavy artillery support. Fixing a 15-cm gun on to an already built *Jagdpanzer* 38(t) chassis seemed a very sensible idea. For this prototype, a *Bergepanzerwagen* 38(t) was used.

This self-propelled artillery gun was officially designated the 15-cm sIG 33/2 (Sf) *auf Jagdpanzer* 38(t). The long version of the name is 15-cm *schweres Infanteriegeschütz* 33/2 *Selbstfahrlafette auf Jagdpanzer* 38(t).

Although the word '*Jagdpanzer*' is used in the name, this vehicle was not a tank hunter. The term '*Jagdpanzer*' is a reference to the donor vehicle chassis used to mount the 15-cm heavy infantry field howitzer. This vehicle is a self-propelled mobile artillery gun. The words '*schweres Infanteriegeschütz*' translates to heavy infantry gun.

It is often incorrectly called the Sturmpanzer 38(t). A *Sturmpanzer* is a heavily armoured assault tank. Its crew are protected in a fully armoured box that is enclosed on all four sides and has an armoured roof. The front armour of this vehicle is thick to enable it to get close to enemy strong points before firing its weapon. This description bears no resemblance to the function and appearance of this vehicle.

Were They Actually Built?

Most history books, museums, and websites state that thirty of these *Jagdpanzer* 38(t) variant SPGs were built. The only known photographs of the 15-cm sIG 33/2 (Sf) *auf Jagdpanzer* 38(t) are of the prototype at the factory. Why are there no operational or trials photographs of this vehicle if thirty were built?

The official order requesting six to be built by converting existing *Jagdpanzer* 38(t) chassis and a further twenty-four to be built on new chassis by December 1944 exists. No documentation has yet been found to confirm that the order had been completed or delivered.

If they were built, it is assumed that they were issued to heavy infantry gun companies of armoured infantry regiments on the Eastern Front, but the exact ones have so far not been identified. No regimental records have been found showing receipt and deployment of these weapons. There are no Soviet photographs of captured or knocked out wrecks of this type of vehicle. None have so far been identified in photographs of Allied German military vehicle scrap yards in Western Europe.

At present, it cannot be confirmed that thirty 15-cm sIG 33/2 (Sf) *auf Jagdpanzer* 38(t) were built even though an official order was issued authorising their manufacture. It can only be confirmed that a prototype was constructed.

Walter J. Spielberger's book *Panzers 35t and 38t and Their Variants 1920–1945* on pages 171–172 states that an additional thirty were built between December 1944 and February 1945 and used as a support vehicle for the *Panzergrenadiers*. Six were converted from existing *Jagdpanzer* 38(t) chassis and twenty-four newly built.

Unfinished *Jagdpanzer* 38(t) Hetzers would come from a holding area off the production line. It was not uncommon for production completion to be held up waiting for the main gun to be finished in a different or separate part of the factory.

The 15-cm *Steilgranate* 42 Projectile

A short-range blast effect weapon was trialled. It was called the 15-cm *Steilgranate* 42 (Stick grenade model 1941). It looked like a small fat space rocket with three stabilisation fins and was used like a very large rifle grenade. There was a large driving rod tube behind the warhead, filled with propellant powder. It was loaded into the end of the gun barrel of the 15-cm sIG 33/2 (Sf) *auf Jagdpanzer* 38(t) SPG. The projectile always remained outside the gun barrel. The driving rod would separate from the warhead when it was fired, it was the German equivalent of the British Petard spigot mortar fitted on their Churchill AVRE tanks on D-Day 6 June 1944.

It had a maximum range of 1,000 m and had a warhead of 27 kg (60 lb) of amatol explosive. It was a demolition charge designed to destroy roadblocks, bunkers, pillboxes, and other concrete, brick, or earthen obstacles. It was used to clear minefields and barbed wire. Unlike other *Steilgranate* ammunition, this version was not intended to be used to knock out tanks. It weighed 90 kg (200 lb) and was difficult to load. It had a more powerful destructive force than the standard 15-cm HE shell, having nearly seven times more explosive.

Surviving SPGs

There are no surviving examples of the 15-cm sIG 33/2 (Sf) *auf Jagdpanzer* 38(t).

Specifications

Length: 4.83 m (without gun) (15 feet 10 inches)
Width: 2.59 m (8 feet 6 inches)
Height: 1.87 m (6 feet 1 inch)
Total weight, battle ready: 15.75 tonnes (15.50 tons)
Armament: 15-cm (5.9-inch) *schweres Infanterie Geschütz* 33 howitzer
Armour: 8–60 mm (0.3–2.36 inches)
Crew: Four (driver, commander, gunner, and loader)
Propulsion: Praga six-cylinder gasoline/petrol 160-hp engine
Speed: 42 km/h (26 mph)
Suspension: Torsion bar
Range: 177 km (110 miles)
Total production: One (or thirty)

The prototype 15-cm *schweres Infanteriegeschütz* 33/2 *Selbstfahrlafette auf Jagdpanzer* 38(t) artillery self-propelled gun in *dunkelgelb* livery. (*David Bocquelet*)

Fictional whitewashed livery of a 15-cm sIG 33/2 (Sf) *auf Jagdpanzer* 38(t) artillery self-propelled gun, ready for winter active service on the Eastern Front in early 1945. (*David Bocquelet*)

Fictional livery of a 15-cm sIG 33/2 (Sf) *auf Jagdpanzer* 38(t) artillery self-propelled gun ready for spring deployment on the Eastern Front in 1945 in *dunkelgelb*, green, and brown livery. (*David Bocquelet*)

19

10.5-cm leFH 18/40 L/28 (Sf) *auf Geschüetzwagfen* III/IV Hummel-Wespe

This armoured fighting vehicle was a merger of two different vehicle designs. There were two main types of self-propelled guns in the German Army during the Second World War. One was fitted with an anti-tank gun and the other with an artillery howitzer, like the Hummel.

There is only one photograph of a Hummel self-propelled gun chassis and body fitted with an artillery 10.5-cm leFH 18/40 L/28 howitzer (usually fitted to on the Wespe Artillery SPG) rather than the normal 15-cm sFH 18/1 L/30 howitzer. It was officially called the Hummel-Wespe. This name was used on the Stahlindustrie construction company's documentation. It was also known as the 10.5-cm leFH 18/40 (Sf) *auf Geschüetzwagfen* III/IV. The photograph was taken after the Second World War during the winter of 1945–46 at Košťaty near Teplice, Czechoslovakia, close to the factory that assembled the vehicle.

The Hummel SPG used an extended tank chassis called the *Geschüetzwagfen* III/IV to mount the 15-cm sFH 18/1 L/30 howitzer. The engine was moved from the rear of the tank to the centre of the vehicle to make room for the gun and the armoured fighting compartment at the back of the SPG.

The 10.5-cm leFH 18/2 (Sf) *auf Geschützwagen* II *Wespe* (Sd.Kfz.124) artillery SPG used a Panzer II tank chassis. Production began in February 1943 and ceased in June 1944, when the main factory in Warsaw, Poland was captured by the Red Army. The German Army *Panzer-Artillerie* batteries still needed more self-propelled guns that could keep up with the panzer divisions, fitted with 10.5-cm leFH 18/40 howitzers.

The *Geschützwagen* III/IV was still in production and being used for the Nashorn 88-mm anti-tank self-propelled gun as well as the 15-cm Hummel SPG. A decision was made to mount the 10.5-cm leFH 18 howitzers gun used on the Wespe SPG on to the *Geschüetzwagen* III/IV chassis.

The German armaments company Deutsche Eisenwerke (DEW) were constructing *Geschüetzwagen* III/IV chassis at their assembly plant in Duisburg, Germany. Allied bombing was making production difficult. It was moved to the DEW plant Werke, Teplitz-Schönau in Czechoslovakia (now known as Teplice, Czech Republic). The building of armoured vehicles for the German Army continued until the end of the war in May 1945.

The open-topped back design of these self-propelled guns had a number of advantages. The elevated commander's position when standing in the crew compartment, behind the protective armoured shield, meant that he had an unobstructed view on all sides. If there was the threat of enemy small arms fire, then the crew could use a twin lens rangefinder telescope that could peak over the top of the armoured casement.

There was enough room for the crew to be transported towards the battlefield while protected from small arms fire and shell shrapnel. The vehicle had good mobility and could follow the infantry almost anywhere. The gun was quicker to get ready for action and fire on targets than towed artillery guns. Putting the 10.5-cm leFH 18/40 howitzer on top of a tank chassis was a more efficient use of manpower from the traditional form of German artillery battery transportation.

High-explosive shells came in two parts. The explosive shell was loaded first, followed by the variable charge canister. This meant that the Hummel/Wespe could only carry eighteen rounds of HE. It could fire armour-piercing rounds but they were only effective at short ranges and used in self-defence. The Hummel/Wespe was not meant to be on the battlefield front line. It was a support vehicle that provided artillery support from behind the infantry and tanks.

The Hummel/Wespe was designed to be operated by a crew of six: commander, driver, and four gunners. They were protected by an enclosed high silhouette armoured fighting compartment. Although it was open-topped, the crew were issued with a thick canvas tarpaulin cover that could be used in bad weather.

In front of the driver, a metal wire grid was fixed into position to aid the driver manoeuvring the vehicle in the correct fire position. These were designed to prevent grenades and mines being thrown into the vehicle as it moved through towns and cities.

A metal-louvred cover ventilated the engine, but many later versions were fitted with an angled shield that opened upward. On the photograph of the Hummel/Wespe, the metal-louvred engine vent cannot be seen on the sides of the vehicle. It looks like it has been fitted with one of the armoured angled shields.

Three aiming stake poles would have been carried in brackets below the rear door. The gunner would use a large ZE 34 sight. The top lens aperture would point to the rear of the vehicle. The gunner used this aperture of the sight to locate the aiming sticks that a member of the crew had pounded into the ground at the rear at a known bearing from the vehicle, having used a compass (compasses did not work inside a metal vehicle in 1943). By lining up the red and white fire aiming stake, subtracting 180 degrees, he would be able to work out the correct bearing the gun barrel is pointing towards.

The upper fighting compartment superstructure walls were constructed using 10-mm (0.39-inch) thick E11 chrome-silicon armour plates hardened to 153 kg/mm^2 for protection against shell fragments. The 30-mm (1.18-inch) thick front hull was made using face-hardened FA32 armour plates. The rest of the hull was made out of cheaper rolled SM-Stahl (carbon steel) that was hardened to 75–90 kg/mm^2. It took 20-mm (0.78-inch) thick plates of SM-Stahl to provide equivalent protection against penetration by SmK (7.92-mm AP bullets) as 14.5 mm (0.57 inches) of E11 armour plate.

The Anti-Grenade Screens

Another unusual feature visible in the surviving photo of the Hummel/Wespe is the hand grenade protection screens that were added over the open fighting compartment on a metal hinged frame.

The *Geschützwagen* III/IV chassis

The powerful 10.5-cm leFH 18/40 light field howitzer was mounted on a specially designed Alkett/Rheinmetall-Borsig lengthened German tank chassis called the *Geschützwagen* III/IV. Components were adopted from both the Panzer III and Panzer IV tank chassis. The more robust final drive wheels, front drive wheels, and steering units plus the Zahnradfabrik SSG 77 transmission gearbox were adopted from the Panzer III Ausf. J.

The Maybach HL 120 TRM engine with its cooling system, the suspension, and idler with track tension adjustment were adopted from the Panzer IV. The engine was moved from the rear of the tank to the centre of the vehicle to make room for the gun and the armoured fighting compartment at the back of the SPG.

On early versions of the *Geschützwagen* III/IV chassis, the front top of the hull had sloping armour with a raised armoured compartment for the driver on the left of the vehicle. The front hull superstructure and driver's armoured compartment were redesigned in early 1944 and enlarged, covering the whole width of the vehicle. The radio operator and driver now had more space to work in. This design was used on all Hummel/Wespe artillery SPGs.

The exhaust system was also changed on the later model. It was moved from the original location below the rear double doors. The exhaust silencers were dropped and the ends of the exhaust pipes were cut at a slant away from the tracks to avoid stirring up additional dust.

The *Geschützwagen* III/IV tank chassis did not have a hull-mounted machine gun. Crews were issued with a single MG 34 or MG 42 machine gun, carried inside the fighting compartment, for self-defence.

The early chassis used the standard 1943 38-cm-wide SK18 track that had three smooth metal pads visible on the front face of the track. In winter, some vehicles were fitted with track width extenders called *Winterketten* (winter track). These triangular pieces of metal were bolted on to the outer edge of the track to extend the width of the track and help the vehicle move across snow and mud by spreading the load over a larger area. They were problematic: they fractured and often fell off. In 1944, vehicles started to be fitted with the wider *Ostketten* (east track) to cope with the conditions found on the Eastern Front. The *Winterketten* extensions made the SK18 tank track 55 cm wide. The one-piece *Ostketten* was 56 cm wide and did not have bits falling off it.

Production Figures

Plans to fit a 10.5-cm LeFH 18/40 light field howitzer to the extended and modified Panzer III/IV tank chassis, as a stop-gap solution to send more 10.5-cm artillery SPGs on the battlefield, were discussed in a meeting on 2 December 1944. The factory was expected to produce forty in February, fifty in March, and eighty in April. A further report documents the demand for a further 250 to be built in June 1945. That report was dated 9 January 1945.

A Stahlindustrie report dated 30 August 1945 stated that one Hummel-Wespe artillery SPG had been built in December 1944, a further nine in January 1945, and a further one before the end of the war, bringing the total to eleven. No German Army documents have been found that show these vehicles entering operational service or being used on the battlefield.

A German document was captured by the Red Army. It showed predicted production numbers of vehicles like the *Jagdpanther*, *Jagdtiger*, *Flakpanzer*, Hummel, and Hummel-Wespe from March 1945 to August 1945. It was translated by Senior Lieutenant Rubinshtein into Russian and kept in the Soviet archives.

On line 345, it shows that planned production of the heavily armed howitzer, the Hummel, was for fifty vehicles: twenty in March and ten each in April, May, and June 1945 with production stopping in July. On line 346, it shows that planned production of the light armed howitzer, the Hummel-Wespe, was for 190 vehicles: twenty in March and thirty each in April, May, and June and forty each in July and August 1945.

There was a side note that an additional ten to twenty new type Hummel/Wespes would be added to the number on April and May. This would have brought the planned production numbers of the 10.5-cm Hummel-Wespe Artillery SPG to 220 vehicles by August 1945. Obviously, that did not happen as the war ended in May 1945. It has been impossible to confirm the total number built. Eight vehicles are known to have survived the war.

Czechoslovakian Army, 1948–54

Surviving Hummel-Wespe self-propelled artillery guns were used by the Czechoslovakian Army after the Second World War. They had twelve, but only eight vehicles underwent renovation and entered service in 1948. Between 1948 and 1949, they were officially called '*Samohybné děla so* 105-mm *húfnicou*'. Between 1949 and 1954, the official designation in army records changed to '105-mm *samohybná húfnica vz.*18/40'.

According to Vojenská história 4/2009 ISSN 1335-3314, VHÚ Bratislava, they were withdrawn from army service in 1954 and presumably scrapped. Czechoslovakian Army records noted the original German production chassis number (Fgst.Nr.) of the eight Hummel-Wespe artillery SPGs that entered their service:

German *Fahrgestellnummer* 84407, date in service 4 May 1949, battery number R114, army registration number 79.651

German *Fahrgestellnummer* 84412, date in service 4 May 1949, battery number R107, army registration number 79.652

German *Fahrgestellnummer* 340003, date in service 4 May 1949, battery number R108, army registration number 79.653

German *Fahrgestellnummer* 84410, date in service 4 May 1949, battery number R3397, army registration number 79.654

German *Fahrgestellnummer* 84422, date in service 20 October 1949, battery number R113, army registration number 79.655

German *Fahrgestellnummer* 84419, date in service 20 October 1949, battery number R109, army registration number 79.656

German *Fahrgestellnummer* 84420, date in service 20 October 1949, battery number R106, army registration number 79.657

German *Fahrgestellnummer* 84421, date in service 20 October 1949, tactical unit number R105, army registration number 79.658

Surviving SPGs

There are no surviving examples of the 10.5-cm leFH 18/40 L/28 (Sf) *auf Geschüetzwagfen* III/IV Hummel-Wespe.

Specifications

Length: 7.17 m (23 feet 5 inches)
Width: 2.97 m (9 feet 7 inches)
Height: 2.81 m (9 feet 2 inches)
Total weight, battle ready: 22 tonnes (21.65 tons)
Crew: Six (commander, driver, and four *Kanoniere* gun crew)
Propulsion: Twelve-cylinder, water-cooled Maybach HL 120 TRM 11.9-litre petrol engine, 265 hp at 2,600 rpm
Fuel capacity: 400 litres
Top road speed: 42 km/h (26 mph)
Operational range (road): 215 km (133 miles)
Main armament: 10.5-cm leFH 18/40 howitzer with eighteen rounds
Secondary armament: 7.92-mm (0.31-inch) MG 34 machine gun
Hull armour: Front 30 mm (1.18 inches); sides and rear 20 mm (0.79 inches)
Superstructure armour: 10 mm (0.39 inches)
Total production: At least eight

10.5-cm leFH 18/2 (Sf) *auf Geschützwagen* III/IV Hummel-Wespe artillery self-propelled gun with a protective anti-grenade wire mesh screen. (*David Bocquelet*)

Above: 15-cm sFH 18/1 (Sf) *auf Geschützwagen* III/IV Hummel artillery self-propelled gun (Sd. Kfz.165) with a protective anti-grenade wire mesh screen. (*David Bocquelet*)

Below: 10.5-cm leFH 18/2 (Sf) *auf Fahrgestell Panzerkampfwagen* II Wespe (Sd.Kfz.124) artillery self-propelled gun. (*David Bocquelet*)

20

17-cm K72 L/50 *und* 21-cm M18/1 (Sf) *auf Geschützwagen* Tiger *Grille* 17/21

On 6 May 1942, the German weapons manufacturer Krupp submitted a proposal for the construction of a new armoured self-propelled gun carriage that used Tiger tank components. While it was based on the Tiger tank chassis design, it was radically altered. It was meant to be able to carry two different guns.

The vehicle was named, in typical German fashion, the *Geschützwagen* Tiger *für* 17-cm *kanone* 72 (Sf) or the *Geschützwagen* Tiger *für* 21-cm *Mörser* 18/1 (Sf), depending on the gun mounted.

Although the German word *Geschützwagen* translates to 'gun vehicle', this is not an accurate description of the concept behind this self-propelled artillery gun; gun carriage would be a better description. Unlike other German self-propelled guns, this vehicle was designed to mount different weapons; it was a modular concept. The vehicle was given the shorter names of *Grille* 17 and *Grille* 21 depending on what weapon was mounted inside the vehicle. The word German *Grille* means 'cricket' and the letter 'e' at the end is pronounced as an 'a'.

The proposal was submitted on 6 May 1942 to the Wa Prüf 4 artillery division of the *Heereswaffenamt* (HWA) (German High Command's centre for technical weapons development). Krupp was authorised to build a single prototype with a completion date of 1 November 1942. The Wa Prüf 4 made a requirement that the vehicle must have the ability to have a 360-degree traverse. They also wanted it to be available for coastal defence work if necessary.

Design and Development

The two guns were too heavy to be mounted in a turret so the Krupp design team had to find another solution. They constructed a large heavy circular base plate that would be carried at the rear of the vehicle and lowered into position when needed. The SPG would then drive on to the metal plate and could swivel on its tracks to point the gun at the target. This was an unusual design feature of this weapon system not seen on any other German vehicle in the Second World War.

Another requirement was that the guns were to be dismountable. This would be achieved by driving backwards towards the base plate, after which the gun could be slid out of the vehicle and mounted on the base plate, allowing it to cover 360 degrees. The reasoning behind this feature was that the *Grille* was also meant to be used in the coastal defence role and this allowed it to fire in any direction. This requirement was dropped in 1944 under the orders of Heinrich Himmler.

Design and mechanical problems were also encountered with the Tiger chassis, engine, and transmission. It did not help that the winner of the competition between the Porsche and the Henschel designs had not yet been decided, and they had vastly different drivetrain arrangements.

Initially, the vehicle was meant to have a 30-mm armour plate on the front of the chassis and 16 mm on the sides. In November 1942, it was decided to use SM-Stahl (carbon steel) in the construction of this self-propelled gun. Some 50 mm of SM-Stahl carbon steel were used for the front of the vehicle. The sides and rear would have 30-mm SM-Stahl carbon steel. This added to the weight of the vehicle. There were a number of delays in the project. The original 1 November 1942 completion date passed without a prototype being finished.

When the Panzer VI Ausf. B Tiger II (Sd.Kfz.182) heavy tanks started rolling out of the factory doors, Krupp decided to use the Tiger II engine, suspension, steering, and transmission instead of Tiger I parts. These components were not ready for delivery until January 1944. This delayed the estimated final construction of the prototype until the summer of 1944.

On 25 September 1944, *Reichsminister* Albert Speer ordered a demonstration for Adolf Hitler to take place as soon as the vehicle was completed, now planned for the end of the year. Serial production was to then start at the rate of two per month.

A Troubled End

More delays occurred when the Allied air force bombed Krupp's manufacturing plant in Essen. Construction work on the prototype was no longer viable at this location. On 7 December 1944, Krupp reported that the chassis was ready to be loaded on to a flat back railway wagon for transportation from Essen to the Henschel *Panzerversuchsstation* 96, Haustenbeck near Paderborn. It was recorded as being at this establishment on documents dated 22 December 1944 but missing many of the components needed for the completion of the project including the cooling and fuel system, Gg 24/800/300 tracks, and hardened road wheel arms.

The *Grille* 17/21 prototype was still in an unfinished condition when the German High Command ordered all future work on the program to be stopped. The situation in the first quarter of 1945 was such that in their view, there would not be any significant advantage in the completion of the project. Resources were limited and they had to be channelled to more important weapon production lines.

In 1945, the US 3rd Army captured the Henschel *Panzerversuchsstation*, Haustenbeck Ordinance proving and tank testing ground in northern Germany, 50 km south-west of Hanover. A selection of German heavy tanks and self-propelled guns were found in working condition. A few prototype vehicles were discovered that never entered production. These included a partially assembled *Geschützwagen* Tiger *für* 17-cm *Kanone* 72 chassis and near by a 17-cm *Kanone* 72. They did not find a second chassis or a 21-cm *Mörser*.

Armament

It was envisaged that two different guns could be mounted in the vehicle: the 17-cm *Kanone* K72 (Sf) L/50 or the 21-cm *Mörser* 18/1 L/31. These two weapons were chosen because they used the same gun carriage and recoil system. The fittings would be the same when the guns were mounted within the self-propelled gun superstructure. The guns would have had a traverse of 5 degrees left and right from a fixed position. The gun sight was a ZE 34 with Rblf36. Both vehicles would have to be supported by a number of ammunition carrying vehicles.

The *Grille* 17 would have carried five rounds plus propellant on board when equipped with a 17-cm *Kanone* K72 (Sf) L/50 gun. It could fire two types of shells: the 68-kg *Sprenggranate* (HE) with 29.15 kg of propellant and a range of 28,000 meters and the 62.8-kg *Sprenggranate* (HE) with 30.5 kg of propellant and a range of 29,600 meters.

The 21-cm *Mörser* 18/1 L/31 was already in production and use by the Germans when the project started. It was produced to replace the much older 21-cm *Mörser* 16. The 21-cm *Mörser* 18 replaced the 21-cm *Mörser* 16 in front-line service around 1940 with the older gun being relegated to secondary theatres and training units. More than 711 21-cm *Mörser* 18 guns were produced from 1939–45. The *Grille* 21 would have carried three rounds plus propellant on board when equipped with this gun. It could fire a 113-kg *Sprenggranate* (HE) shell with 15.7 kg of propellant to a range of up to 16,700 meters.

In January 1945, plans were made to mount a 30.5-cm calibre smooth bore mortar with fin-stabilised projectiles, due to the concerns over the length of time it took to produce artillery barrels for the two other guns. Krupp and Skoda both competed on this project with Skoda producing a 30.5-cm GrW L/16 prototype by April 1945.

As a side note to armament, in 1945, Kurt Arnoldt, chief engineer at Henschel, said in a 1945 interview that the 21-cm gun produced too much recoil for the chassis as designed. Making it impossible to fire from the chassis. The 17-cm muzzle break is based upon a design by Solothurn design. Ammunition would have been stowed both in the vehicle and in wicker baskets on the side of the vehicle as well as in a following 18-ton semi-tracked vehicles (half-track). The gunnery sights also allow for direct firing of the gun on close-range targets.

Crew Members

According to a British intelligence report from 1945, the *Grille* designs were meant to have a crew of eight, composed of a driver, a commander, and six gun crewmen. The loading of the two-part ammunition was meant to be done manually. It is stated in a 1945 interview with Kurt Arnoldt, the chief engineer for Henschel, that the additional crew and ammunition would be transported in a half-tracked vehicle.

Mobility

A Tiger II tank's Maybach HL 230 P30 V-12 water-cooled petrol 690-hp engine was ordered along with a Maybach OG 40 12 16 B gearbox with eight forward and four reverse gears. Henschel made the L 801 steering unit. The engine was mid-mounted in the chassis to allow for the maximum amount of room for the large gun and space for the crew.

As rubber was difficult to obtain in 1944, the wheels were 80-cm all-steel tires. However, the 1945 interview with Kurt Arnoldt suggests that these would be rubber-rimmed roadwheels. Photos do not confirm this. It was fitted with Gg 24-800/300 Tiger II tank track for travelling cross country, but this would have been replaced with smaller width Gg 24-600/300 Panther II tank track if transportation by rail was required.

Surviving SPG

There are no surviving examples of the *Grille* 17/21 artillery SPG.

Specifications

Length *Grille* 17: 13 m (42 feet 8 inches)
Length *Grille* 21: 11 m (36 feet 1 inch)
Length without gun: 9.50 m (31 feet 2 inches)
Width *Grille* 17: 3.27 m (10 feet 9 inches)
Width *Grille* 21: 3.59 m (11 feet 8 inches)
Height: 3.15 m (10 feet 4 inches)
Propulsion: Maybach HL 230 P30 V-12 23 litre water-cooled petrol 650-hp engine
Main armament: 17-cm K72 L/50 or 21-cm M18/1 L 31 mortar
Gun elevation: -2 and +48 degrees
Mortar elevation: +68 degrees
Crew: Eight (commander, driver, and six gunners)
Weight: 60 tonnes (59 tons)
Production front armour: 50-mm carbon steel
Production sides and rear armour: 30-mm carbon steel
Prototype front armour: 30-mm armour plate
Prototype sides and rear armour: 16-mm armour plate
Front armour: 30-mm armour plate
Sides and rear armour: 16-mm armour plate
Speed on/off road: 45 km/18 km (28 mph/11 mph)
Range on/off road: 250 km/125 km (155 miles/78 miles)

The Allied soldiers found the prototype Grille 17/21 artillery self-propelled gun carriage at the Henschel Panzerversuchsstation, Haustenbeck Ordinance proving and tank testing ground in northern Germany without a gun fitted. (*David Bocquelet*)

Above: The *Geschützwagen* Tiger *für* 17-cm *kanone* 72 (Sf.) artillery self-propelled gun, *Grille* 17 SPG with rear turning platform fixed in the up position and gun lock deployed for transportation. (*David Bocquelet*)

Below: Geschützwagen Tiger *für* 21-cm *Mörser* 18/1 (Sf.) artillery self-propelled mortar, *Grille* 21. When the Allies captured the Henschel tank testing depot, some of the vehicles were just painted in red-oxide prima. The guns were not mounted in the chassis. The rear turning platform is not present in this illustration. They were removed for firing and placed on the floor. (*David Bocquelet*)

21

12.2-cm FK (r) *auf* GW Lorraine Schlepper(f)

During the Second World War, Polish, Soviet, German, and British armed forces used armoured trains. Germany had twenty-one armoured trains in 1942, twenty-nine in 1943, forty-four in 1944, and fifty-five in 1945. The Soviets had a lot more, including captured Polish armoured trains.

The locomotive would be covered in protective armour plate and pull artillery wagons fitted with howitzers, anti-aircraft wagons bristling with flak guns for self-defence, anti-tank wagons with tank turrets mounted on top of an armoured coach, command and assault wagons to carry troops plus at the back a flatbed tank transporter with ramps.

A variety of different tanks were carried on these trains at the rear. The tank could dismount from the train when needed and take the battle to the enemy by outflanking them or using direct assault under the cover of supporting fire from the train. Sometimes, as in this case, a self-propelled artillery gun was used instead of a tank.

The 12.2-cm FK (r) *auf* GW Lorraine Schlepper(f) self-propelled artillery gun was built around a captured French Army Lorraine 37L armoured tracked supply vehicle that had a captured Soviet 12.2-cm gun mounted on top of it. An armoured open-topped superstructure was built around the gun and gun crew to protect them from small arms fire and shell shrapnel.

This vehicle was used on the PZ32 German armoured train in France. It would be carried at the end of the train on a flatbed tank transporter wagon that had ramps. The driver would reverse the 12.2-cm FK (r) *auf Geschützwagen* Lorraine Schlepper(f) up the ramps and on to the wagon.

Many observers have claimed it was used as an anti-aircraft gun as in many photographs, the gun barrel was pointed up into the sky. It was never used to try and shoot down aircraft. It was a long-range artillery gun designed to deliver high-explosive shells great distances.

The gun was fixed into position. It was not in a turret that could turn to face the enemy. It had a limited traverse left and right. It was envisaged that when the train got to the battlefield the 12.2-cm FK (r) *auf* GW Lorraine Schlepper(f) SPG would be driven off the train, down the rear ramps, and moved to face the direction of the target.

It could be used in direct fire mode at targets the crew could see, but more commonly it was used for indirect fire at targets plotted on a map. It was not designed to be on the front line or engage in combat with tanks. The vehicle's armour was not thick enough to prevent AP shells from enemy tanks penetrating its armour. Most targets would have been given to the crew as map grid references by forward observation officers or infantry units under attack. Quite often, the gun crew could not see where their shells landed, as the target was so far away. They would have to rely on the forward observer to tell them if adjustments had to be made.

New Build or Battlefield Conversion?

The origins of the 12.2-cm FK (r) *auf* GW Lorraine Schlepper(f) self-propelled artillery gun are not known. Was a captured Soviet 12.2-cm (4.8-inch) howitzer M1938 (M-30) used to replace the gun in a Marder I Lorraine Schlepper(f) anti-tank self-propelled gun or a 10.5-cm leFH-18/40 *auf Geschützwagen* Lorraine Schlepper(f) self-propelled artillery gun? Was this vehicle a one-off new build?

A company called Alkett, based near Berlin, built the first 10.5-cm leFH-18/40 *auf Geschützwagen* Lorraine Schlepper(f) self-propelled artillery guns in 1942. German Army engineer Major Alfred Becker worked with the company to create these conversions.

The next year, he was in Normandy at the head of a *Baukommando*, a construction command unit. Becker's men, engineers, and mechanics converted more Lorraine 37L tractors into self-propelled artillery guns by fixing 10.5-cm leFH-18/40 howitzers on to the top of these vehicles. The design of the fighting compartment was slightly different on each version. By comparing photographs of the Becker-and-Alkett-built SPG with the 12.2-cm FK (r) *auf* GW Lorraine Schlepper(f), it can be seen that major work was done on the top section of the upper armour.

It looks like the original vehicle was an Alkett-built 10.5-cm leFH 18/40 *auf* GW Lorraine Schlepper(f), as the upper and lower armour joints of the fighting compartment are low down. That joint on the Becker-built vehicles was much higher.

However, where the fighting compartment upper side armour plates meet the front plates, the joint angle is forward not backward as on the Alkett build.

Was it a converted Marder I that was also built on a Lorraine 37L chassis? There was no forward gun lock on this SPG as there was on the Marder I. The angle of the joint between the upper front and side armour was the same unlike on the 10.5-cm leFH 18/40 *auf* GW Lorraine Schlepper(f).

Walter J. Spielberger, author of *Beute-Kraftfahrzeuge und panzer der deutschen Wehrmacht*, had the benefit of an interview with Major Becker after the war and access to all the original documents. There is only one small paragraph about this vehicle in the book.

It does say that only one vehicle was built with a 12.2-cm *Kanone* (r) on a Lorraine-*Schlepper* tractor. It does not say that the Soviet gun replaced a 10.5-cm leFH 18 gun or a 7.5-cm PaK 40 L/46 gun. This vehicle could have been a new build rather than a battlefield conversion of an existing self-propelled gun.

The 12.2-cm Howitzer M1938 (M-30)

Between 1939 and 1955, Soviet factories produced 19,266 of these artillery howitzers. They were developed by the design bureau of Motovilikha Plants, headed by F. F. Petrov, in the late 1930s. It was used as a towed divisional artillery piece during the Second World War.

The Finish and German armies reused captured guns. The German Army gave them the designation 12.2-cm sFH 396(r) heavy field howitzers. Germany began mass production of 12.2-cm (4.8-inch) ammunition for these and other captured howitzers, producing 424,000 shells in 1943, 696,700 in 1944, and 133,000 in 1945.

The German 10.5-cm leFH 18/40 light field howitzer had a muzzle velocity of 540 m/s, elevation of 45 degrees and a range of 12,325 m. The Soviet M-30 12.2-cm gun had a muzzle velocity of 515 m/s, elevation of 49 degrees, and a range of 11,720 m. Both guns were similar but the German howitzer had a less powerful high-explosive shell and its smaller maximum elevation made it less effective against dug-in troops.

The howitzer was designed to fire high-explosive and smoke shells. In May 1943, a 12.2-cm high-explosive anti-tank round became available. It only had a range of 2,000 m and the muzzle velocity was reduced to 335 m/s. This was meant for self-defence. This gun was an artillery howitzer, not a high-velocity anti-tank gun.

The Train (1964 Film)

The Train was a Hollywood movie starring Burt Lancaster and the PZ32 armoured train that was captured in St Berain, France. The story is set in 1944. A German colonel loads a train with French art treasures to send to Germany. The Resistance must stop it without damaging the cargo. The 12.2-cm FK (r) *auf* GW Lorraine Schlepper(f) self-propelled gun can be seen at the rear of the train. It is not known what happened to the train or the SPG after filming had finished.

Specifications

Length: 4.22 m (without gun) (13 feet 10 inches)
Width: 1.57 m (5 feet 2 inches)
Height: 2 m (6 feet 7 inches)
Total weight, battle ready: 7.7 tonnes
Crew: Four (commander, driver, and two *Kanoniere* gun crew)
Propulsion: Type 135 Delahaye six-cylinder inline petrol engine, 70 hp at 2,800 rpm
Fuel capacity: 144 litres
Top road speed: 35 km/h (22 mph)
Operational range (road): 137 km (85 miles)
Main armament: Soviet 12.2-cm (4.8 inches) howitzer M1938 (M-30) (German designation: 12.2-cm sFH 396(r))
Hull armour: Front 9 mm (0.35 inches), cast nose 12 mm (0.47 inches), sides 9 mm (0.35 inches), rear 9 mm (0.35 inches)
Superstructure armour: 10 mm (0.39 inches)
Total production: One

The 12.2-cm FK (r) *auf Geschützwagen* Lorraine Schlepper(f) self-propelled artillery gun carried on the rear wagon of the German PZ32 Armoured train.

Compare the design of the superstructure around the gun on this 10.5-cm leFH-18/40 *auf Geschützwagen* Lorraine Schlepper(f) self-propelled artillery gun. (*Alkett version*)

Compare the design of the superstructure around the gun on this 10.5-cm leFH-18/40 *auf Geschützwagen* Lorraine Schlepper(f) self-propelled artillery gun. (*Baukommando Becker version*)

Compare the design of the superstructure around the gun on this 7.5-cm PaK 40/1 *auf Geschützwagen* Lorraine Schlepper(f) Sd.Kfz.135 (Normandy, 1944 Marder I).

Bibliography

Books

Bingham, J., *French Infantry Tanks: part I,*. (Profile Publications)
Chamberlain, P., and Doyle, H. L., *Profile AFV Weapons 55 German Self-Propelled Weapons,* (Profile Publications); *German Army S.P. Weapons 1939-45 Part 2. Handbook No1*, (Bellona)
Chamberlain, P., and Gander, T., *Wespe (SD.KFZ.124),* (Nuts and Bolts); *Weapons of the Third Reich,* (Doublesday); *Small arms, artillery and special weapons of the Third Reich,* (MacDonald and Jane's)
Chamberlain, P., Jentz, T.L., and Doyle, H.L., *Encyclopedia of German tanks of WWII*, (Arms and Armour Press)
de Sisto, F.V., *German Artillery at War 1939-45 vol.1,* (Concord Publication Co)
Englemann, J., *German self-propelled artillery in WWII and other 150mm self-propelled guns,* (Schiffer Publishing Ltd); *German self-propelled artillery in WWII: Wespe Vol 61,* (Schiffer Publishing Ltd); *German Heavy Field Artillery 1934–1945* (Schiffer Publishing Ltd)
Feist, U., *Nashorn, Hummel, Brumbär in Action* (Podzun-Verlag)
Hinze, R., *Die 19. Panzerdivision 1939–45* (Podzun-Pallas)
Hogg, I.V., *German Artillery of WW2* (Pen & Sword)
Jentz, T. L., Doyle, H. L., and Badrocke, M., *Jagdpanzer 38 'Hetzer' 1944–45* (Osprey Publishing)
Jentz, T.L., and Doyle, H.L., *Panzer Tracts No.7-3, Panzerjägers* (Panzer Tracts); *Panzer Tracts No. 10 Artillerie Selbstfahrlafetten,* (Panzer Tracts); *Panzer Tracts No. 10-1 Artillerie Selbstfahrlafetten,* (Panzer Tracts)
Ledwoch, J., *No. 233 Wespe—Tank Power Vol. XVII* (Wydawnictwo Militaria).
Lüdekes, A., *Beutepanzer der Wehrmacht* (Typenkompass).
Parada, G., Suliga, M., and Hryniewicki, W., *Wespe Sd.Kfz 124* (Kagero)
Quarrie, B., *The Ardennes Offensive: V Panzer Armies—Central Sector* (Osprey Publishing)
Rottman, G., *German self-propelled guns* (Concord Publication Co)
Scheibert, H., and Regenberg, W., *Beutepanzer unterm Balkenkreuz* (Wölfersheim-Berstadt)
Spielberger, W. J., *Beute-Kraftfahrzeuge und panzer der deutschen Wehrmacht* (Motorbuch); *Die Panzerkampfwagen I und II und ihre Abarten*, (Motorbuch); *Panzer-35t/38t* (Motorbuch)
Zaloga, S. J., *French Tanks of World War II* (Osprey Publishing)
Zetterling, N., *Normandy 1944: German military organization, combat power and effectiveness* (Fedorowicz)

Article

Fleische, W., '*Die Feldartillerie des Heeres in der Panzerabwehr* 1939–1945', *Militärgeschichte* issue 1/1994, pp. 9–15 (Berlin; Deutscher Militärverlag 1994)

Original Documents

Catalog of Enemy Ordnance, US Office of Chief of Ordnance, 1945

McNair, L. J., Artillery Firing, Lectures to the staff and line classes General Service Schools, Fort Leavenworth, Kansas October 1919.

The War Office, Handbook of Enemy Ammunition Pamphlet No 15—24 May 1945.

SHAEF, Restricted July 1944—Allied Expeditionary Force—German Guns—Brief notes and range tables for allied gunners. SHAEF/16527/2A/GCT.

SHAEF, Allied Expeditionary Force German Guns—Brief Notes and Range Tables for Allied Gunners—SHAEF/16527/2A/GCT July 1944

US Office of Chief of Ordnance, 1945 Catalog of Enemy Ordnance

US War Department, German Infantry Weapons—US Military Intelligence Service, Special Series No. 14, May 25, 1943.

US Army Council. Handbook of Enemy Ammunition Pamphlet No.15. German Ammunition Markings and Nomenclature.

Vojenská história 4/2009 ISSN 1335-3314, VHÚ Bratislava

Internet sources

tank-hunter.com

tankmuseum.org

tanks-encyclopedia.com

the.shadock.free.fr/Surviving_Panzers.html

armada.vojenstvi.cz/povalecna/ctenari/20.htm

warspot.ru